T0135790

Experimental and Mathematical Analysis of Regulatory Networks in T-helper Lymphocytes

Edda G. Schulz

DISSERTATION

im Fach Biophysik

an der

Humboldt-Universität zu Berlin

Bibliografische Information der Deutschen Nationalbibliothek

Die Deutsche Nationalbibliothek verzeichnet diese Publikation in der Deutschen Nationalbibliografie; detaillierte bibliografische Daten sind im Internet über http://dnb.d-nb.de abrufbar.

ISBN 978-3-8325-2499-9

Logos Verlag Berlin GmbH
Comeniushof, Gubener Str. 47,
10243 Berlin
Tel.: +49 (0)30 42 85 10 90
Fax: +49 (0)30 42 85 10 92
INTERNET: http://www.logos-verlag.de

Contents

Part I

The gene-regulatory network governing Th1 differentiation

Chapter 1

Introduction

In the immune system many cell types work together to defend an organism against invading pathogens. Each cell type fulfills distinct tasks, such as the recognition of a certain kind of pathogen, or its elimination. Communication between the immune cells is mediated by cell-contact dependent mechanisms and by soluble messenger molecules. A specific combination of signals activates a cell type dependent genetic program, which controls effector mechanisms, such as the production of antibodies. In a few highly evolved cell types such a genetic program can drive a differentiation process, resulting in stable changes of the cell phenotype.

To advance our understanding of how this complex system functions on the molecular level, one such genetic program is analyzed in detail in the first part of this thesis, namely the transcriptional network that is mediating the differentiation of naïve T-helper cells to so-called type 1 T-helper cells (Th1). Th1 cells plays a critical role in defense against intracellular pathogens and during autoimmune diseases. In the following sections, the role of Th-cells in the immune system and our current knowledge about the Th1 differentiation program are summarized.

1.1 The role of T-helper lymphocytes in the immune system

T-helper lymphocytes are part of the adaptive immune system in higher vertebrates. The adaptive immune system has evolved on top of the innate immune system to allow a more specific and a more powerful defense against pathogens (Litman et al., 2005). The innate immune system relies on a limited number of pattern-recognition receptors that detect pathogen-specific molecules. By contrast, the T- and B-cells of the adaptive immune system have developed complex mechanisms to discriminate between self and unknown non-self. Through somatic DNA-rearrangements Th-cells can express a large repertoire of T-cell receptors (TCR) with distinct antigen-specificities. During T-cell development in the thymus, Th-cells reactive to non-self molecules are selected and then circulate the body searching for their cognate antigen (Murphy et al., 2008).

However, Th-cells do neither take part in initial recognition of the pathogen, nor in its elimination. Instead, they receive information on the invading pathogen and transmit it to effector cells that then clear the infection. First, a professional antigen-presenting cell (APC) takes up the antigen and presents it on its surface. When a pathogen-associated pattern is present, the APC also upregulates co-stimulatory molecules on its surface. The APC can then activate a T-cell specific for the antigen through stimulation of the TCR (and co-stimulation). In this way, the Th-cell is informed about the presence of an infection and can migrate to the site of inflammation. For primary activation of Th-cells usually dendritic cells act as APC. The dendritic cell also senses the type (i.e. intracellular vs extracellular) of pathogen through its pattern-recognition receptors and delivers this information to the T-cell, typically in form of cytokine signals (Fig. 1.1). Depending on the cytokine environment that naïve T-cells experience during primary antigen encounter (priming), they will differentiate and acquire an effector phenotype. Upon re-encounter of the antigen, effector cells are able to express specific sets of cytokines, even in the absence of the initial differentiation signals. Through the cytokines they secrete, effector cells then activate other immune cells to tailor the immune response according to the specific pathogen (Fig. 1.1). Four different effector phenotypes (Th1, Th2, Th17, iTreg) are well established by now, where Th1 and Th2 cells appear to be more stable than Th17 and iTreg cells (Zhou et al., 2009). After clearance of the pathogen, a small fraction of the effector T-cells will become memory cells that provide protection during reinfection with the same pathogen.

1.2 T-helper cell differentiation

During the primary infection, a naïve Th-cell receives several signals: the antigen, stimulating the TCR, co-stimulatory signals, indicating that the APC has experienced an inflamed environment, and cytokine signals that determine the differentiation pathway (Fig. 1.1). During an infection with intracellular pathogens, the APC produces interleukin-12 (IL-12), which induces Th1 differentiation in the T-cell (Murphy, 2006; Zhu and Paul, 2008). A Th1 cell secretes interferon-γ (IFN-γ) and promotes a cell-mediated immune response. By contrast, extracellular pathogens, like helminths, typically trigger differentiation towards Th2 cells (Murphy, 2006; Zhu and Paul,

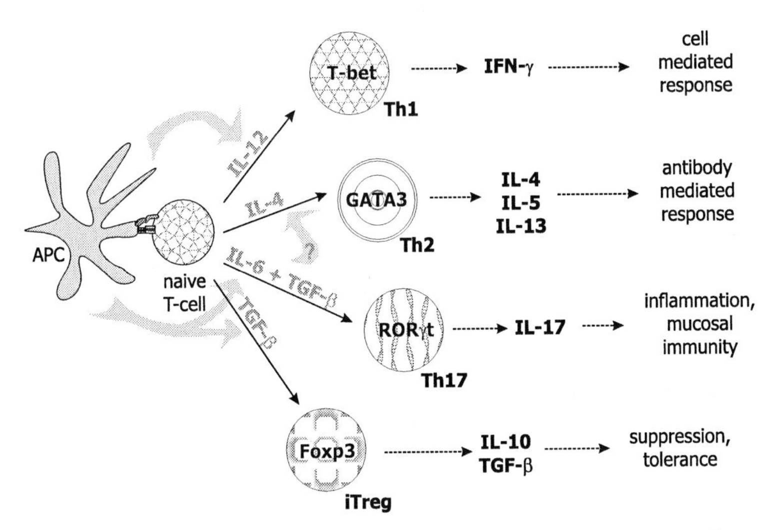

Figure 1.1. T-helper cell differentiation. To differentiate towards an effector phenotype, the naïve T-cell must receive two main signals: The antigen, presented by the APC, which stimulates the TCR, and cytokine signals that control the lineage decision. If the APC secretes IL-12 during primary activation, then Th1 differentiation is induced, driven by the Th1 master transcription factor T-bet. Th1 cells are competent to produce IFN-γ and promote a cell-mediated immune response. If a naïve T-cell is stimulated by IL-4 during its primary activation, the transcription factor GATA3 is induced and the cell adopts a Th2 phenotype, which is able to secrete IL-4, IL-5 and IL-13. These cytokines then promote an antibody mediated immune response by activating B-cells. If primary activation occurs in the presence of TGF-β and IL-6, the transcription factor RORγt is induced and drives differentiation towards the Th17 phenotype. These cells can produce IL-17, promote an inflammatory response and are important for mucosal immunity. Naïve Th-cells can also acquire a suppressive phenotype. In the presence of TGF-β (and in the absence of IL-6), the transcription factor Foxp3 is induced and the cells adopt a regulatory phenotype. These cells can secrete TGF-β and IL-10 and play a role in maintenance of tolerance to self-antigen.

2008). The initial source of the Th2-inducing cytokine, IL-4, is still unclear, but Th2 cells themselves express IL-4 in addition to IL-5 and IL-13. These cytokines promote a humoral, antibody-mediated immune response. In the presence of transforming growth factor β (TGF-β) and IL-6 naïve T-cells become Th17 cells, which play an important role in protection against extracellular bacteria (McGeachy and Cua, 2008). They produce IL-17 and other cytokines that promote tissue inflammation, but their specific mechanisms of action are not yet well understood. Finally, dendritic cells can also present tissue antigens and peptides derived from intestinal microbiota in the absence of infection and inflammation. In this case, the dendritic cells are not activated and therefore produce predominantly TGF-β, which induces iTreg (inducible regulatory T-cells) differentiation (Curotto de Lafaille and Lafaille, 2009). These cells have anti-inflammatory properties and promote tolerance towards self-antigens.

The lineage decision made by naïve Th-cells upon primary antigen encounter is tightly controlled on mainly three different levels:

1. Production of differentiation signals: Cytokine production by dendritic cells is initially regulated by pathogen-derived signals, but later also responds to signals from the T-cells. Th-cell differentiation then depends on the signal secreted by the APC, but is often also controlled by T-cell derived signals, involved in positive feedback regulation.

2. Responsiveness to differentiation signals: Cytokine responsiveness of T-cells is tightly controlled via the expression of cytokine receptors, like the IL-4 receptor and the IL-12 receptor.

3. Expression of transcription factors: Expression and activation of transcription factors that drive differentiation and effector functions are tightly controlled. "T-box expressed in T-cells" (T-bet, Tbx21) promotes Th1 differentiation, "GATA binding protein 3" (GATA3) induces Th2 cells, "RAR-related orphan receptor γt" (RORγt) regulates Th17 cells and "forkhead box P3" (Foxp3) is critical for iTreg differentiation.

All three regulatory levels are used to initiate, reinforce and stabilize the phenotype and to inhibit alternative fates.

1.3 Th1 differentiation

Th1 cells play an important role in host defense against intracellular pathogens, but their erroneous activation does also contribute to the development of autoimmune diseases (Murphy and Reiner, 2002). During viral infections, Th1 cells secrete IFN-γ and thereby induce an anti-viral state in other somatic cells (Schroder et al., 2004). Moreover, IFN-γ promotes production of IgG2a antibodies by B-cells that can neutralize viral particles (Schroder et al., 2004). In many viral infections, cytotoxic T-cells require help from Th-cells to activate dendritic cells (Zhang et al., 2009). Another type of pathogens whose clearance requires a Th1 response are bacteria (e.g. *Tuberculosis, Listeria*) or protozoa (e.g. *Leishmania, Toxoplasma*) that reside inside macrophages (Scott, 1991; Wakil et al., 1998; Yap et al., 2000; Yoshikai, 2006). Through production of IFN-γ (and CD40-CD40L interaction) Th1 cells activate macrophages, thereby enabling them to kill the pathogen (Mosser and Edwards, 2008).

While Th-cells are crucial for the immune response in healthy individuals, they are also responsible for severe damage in many autoimmune diseases (Romagnani, 1996). Then, auto-reactive T- and B-cells target healthy tissue, because the discrimination between self and non-self is perturbed. Activated T-cells together with auto-antibodies destroy the tissue through chronic inflammation. For example, in systemic lupus erythematosus, chronically activated Th1 cells, specific for DNA- and chromatin-derived antigens are critical for disease progression (Cava, 2009). They provide help to B-cells to produce auto-antibodies and they activate macrophages through IFN-γ production. In this disease the Th1/Th2 imbalance is critical for the pathology. Therefore a better understanding of the mechanisms governing Th1 differentiation will also be helpful for the treatment of such diseases.

Cytokines in Th1 differentiation

Two cytokines are known to play central roles in driving Th1 differentiation: IL-12 and IFN-γ. IL-12 is produced by dendritic cells and by macrophages (Hsieh et al., 1993; Macatonia et al., 1995). By contrast, IFN-γ is expressed by the differentiating Th-cells themselves and has a dual role as differentiation signal and effector cytokine (Murphy and Reiner, 2002). Primary IFN-γ acts in an autocrine manner and is required for efficient differentiation (Macatonia et al., 1993; Seder et al., 1993; Schmitt et al., 1994). Differentiated effector Th1 cells then re-express IFN-γ in response to TCR-stimulation in the absence of IL-12 (Ouyang et al., 1999). This memory expression of IFN-γ acts in a paracrine manner, mediating Th1 effector functions. Apart from being expressed by Th-cells, IFN-γ is also produced by other cell types, such as cytotoxic T-cells and natural killer (NK) cells of the innate immune system (Schroder et al., 2004). It depends on the pathogen, whether T-cell derived IFN-γ is sufficient to clear the infection or whether IFN-γ produced by NK cells is also required (Scharton and Scott, 1993; Orange et al., 1995; Macatonia et al., 1993; Wakil et al., 1998).

IL-12 is clearly important to induce Th1 differentiation, since IL-12 deficient mice are defective in mounting a Th1 response (Magram et al., 1996). Moreover, priming of naïve cells in the presence of IL-12 induces Th1 differentiation *in vitro* (Hsieh et al., 1993; Seder et al., 1993). IL-12 has also been suggested to act through selective mechanisms by inducing proliferation in T-cells, but there are conflicting reports on this issue (Perussia et al., 1992; Nishikomori et al., 2000; Heath et al., 2000; Ahn et al., 1998; Mullen et al., 2001). In part, the role of IL-12 as a powerful differentiation signal results from its ability to induce primary IFN-γ expression in the presence of TCR-stimulation. IFN-γ alone, however, is not able to drive Th1 differentiation, but is required for IL-12 induced differentiation (Scott, 1991; Seder et al., 1993; Macatonia et al., 1993; Schmitt et al., 1994). Although it has been known for long that both, IFN-γ and IL-12 are required for Th1 differentiation, their specific contributions are still incompletely understood and were addressed in this study.

Signal transduction in Th1 differentiation

IL-12 as well as IFN-γ signals are mediated by heterodimeric receptors that activate the JAK/STAT signaling pathway (Shuai and Liu, 2003). Upon ligand binding, the receptors dimerize and recruit Janus kinases (JAK), which trans-phosphorylate each other and then the receptor. The phosphorylated receptor provides binding sites for STAT (Signal Transducer and Activator of Transcription) proteins, which are then phosphorylated by the JAK kinases. Phosphorylated STAT proteins dimerize and enter the nucleus, where they act as transcription factors.

The main transcription factor activated by IFN-γ is the STAT1 homodimer. The IFN-γ receptor consists of the R1 and R2 subunits. The R1 chain binds IFN-γ and is constitutively expressed, while expression of the signal-transducing R2 chain is transcriptionally regulated (Bernabei et al., 2001). The R2 chain is rapidly downregulated in a negative feedback mechanism in response to IFN-γ to limit its apoptotic effects on T-cells (Pernis et al., 1995; Bach et al., 1995; Liu and Janeway, 1990). In Fig. 1.2A the regulation of IFN-γ and its downstream effects are summarized.

The IL-12 receptor consists of the β1 and the β2 chains and activates STAT4 (Watford et al., 2004). Regulation of IL-12 signaling is summarized in Fig. 1.2. The β1 chain is also part of the IL-23 receptor that is involved in Th17 differentiation and is therefore not specific for IL-12 signal transduction. Although the β1 chain has been reported to be upregulated on the transcriptional level in response to IFN-γ (Kano et al., 2008), the expression of the β2 chain is rate limiting for downstream signal transduction (Szabo et al., 1997). The IL-12Rβ2 subunit is not expressed on naïve T-cells and it has

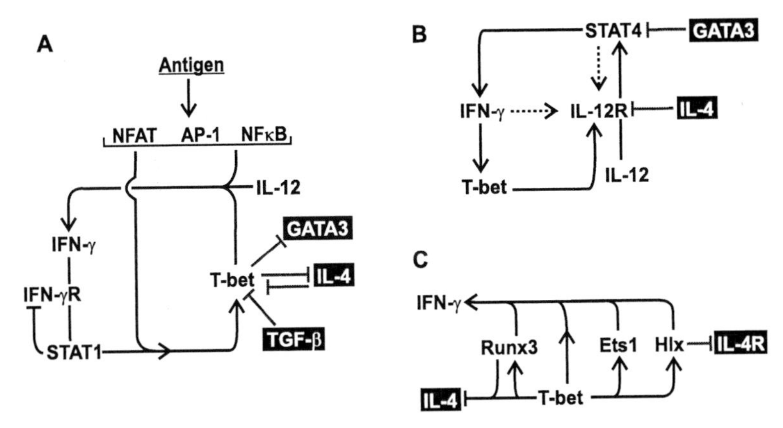

Figure 1.2. Regulatory interactions during Th1 differentiation. **(A)** Schematic representation of the mechanisms controlling expression and effects of IFN-γ. **(B)** Control of IL-12 signaling. **(C)** Feed-forward loops that mediate T-bet dependent imprinting of the *Ifng* locus. Genes mediating alternative differentiation pathways are depicted in white letterss.

been proposed that low-level expression might be induced by TCR-signaling (Szabo et al., 1997; Ouyang et al., 1998). Expression is extinguished rapidly in Th2 cells, because IL-12Rβ2 is repressed by IL-4 (Himmelrich et al., 1998) (Fig. 1.2B, white letters). In Th1 cells IL-12Rβ2 is induced with surprisingly slow kinetics, reaching maximal expression only after several days (Ouyang et al., 1998; Yamane et al., 2000). During Th1 differentiation, IL-12 itself as well as IFN-γ have been identified as positive regulators of IL-12Rβ2 expression (Fig. 1.2B), where IFN-γ seems to be particularly important to override inhibition by IL-4 (Smeltz et al., 2002; Szabo et al., 1997; Rogge et al., 1997). Therefore, IL-12Rβ2 is controlled by a positive feedback loop through IL-12 signaling itself, but also by the second Th1-inducing cytokine, IFN-γ. Thus, one mechanism through which IFN-γ contributes to differentiation is by rendering the cells responsive to IL-12 (Yamane et al., 2000; Hu-Li et al., 1997).

Transcription factors in Th1 differentiation

Since Th1 differentiation is induced by IL-12, STAT4 the transcription factor that is activated downstream of IL-12 should be a critical regulator of differentiation (Bacon et al., 1995; Jacobson et al., 1995). Indeed, *Stat4* deficient cells show defective IFN-γ expression and Th1 differentiation (Kaplan et al., 1996; Thierfelder et al., 1996). However, even when IL-12Rβ2 is constitutively expressed and IL-12 is present, Th1 differentiation cannot always be induced (e.g. in the presence of IL-4), indicating that activation of STAT4 is insufficient to drive differentiation (Nishikomori et al., 2000; Heath et al., 2000).

In contrast to STAT4, the transcription factor T-bet can induce Th1 differentiation even in the absence of IL-12 and has therefore been termed the master regulator of Th1 differentiation (Szabo et al., 2000). When expressed ectopically in cells stimulated with IL-4 in the absence of IL-12, T-bet induces Th1 differentiation, including a high degree of IFN-γ expression and silencing of Th2-specific genes (Szabo et al., 2000). The expression of T-bet is controlled by IFN-γ signaling in synergy with TCR stimuli, as shown schematically in Fig. 1.2A (Afkarian et al., 2002; Lighvani et al., 2001). IFN-γ acts through STAT1, which binds to an enhancer in the *Tbx21* gene ~12kb upstream of the transcriptional start site (Yang et al., 2007). TCR-stimulation by antigen activates three families of transcription factors, namely, "nuclear factor in activated T-cells" (NFAT), "nuclear factor κb" (NF-κb) and "activator protein 1" (AP-1). Which of these transcription factors are required for T-bet induction is still unknown. Since T-bet is not completely absent in *Stat1* deficient T-cells, other pathways seem to contribute to T-bet expression (Lighvani et al., 2001). For example, Notch-signaling and RelB have been implicated in T-bet regulation (Minter et al., 2005; Corn et al., 2005). To inhibit Th1 differentiation in the alternative lineages, T-bet is repressed by IL-4, which induces Th2-differentiation, and by TGF-β, which is involved in Th17 and iTreg induction (Fig. 1.2A, white letters) (Mullen et al., 2001; Gorelik et al., 2002).

A central role of T-bet in Th1 differentiation is supported by the fact that *Tbx21* deficient mice are unable to clear infections that require a Th1 response, such as *Leishmania major* (Szabo et al., 2002). *Tbx21*$^{-/-}$ Th cells show defective IFN-γ production during primary activation and a strong defect in Th1 differentiation, measured as the ability to re-express IFN-γ in a second stimulation in the absence of IL-12 signals (Szabo et al., 2002). Since T-bet activates IFN-γ expression and IFN-γ in turn promotes T-bet induction, a self-reinforcing feedback loop between T-bet and IFN-γ is established (Fig. 1.2A) (Lighvani et al., 2001). T-bet acts on multiple levels to induce Th1 differentiation. It renders cells responsive to IL-12, by inducing the β2 chain of the IL-12 receptor (Fig. 1.2B) (Mullen et al., 2001; Afkarian et al., 2002). T-bet inhibits Th2 differentiation by repressing GATA3 expression and function (Hwang et al., 2005; Usui et al., 2003) and through direct silencing of the *Il4* gene (Szabo et al., 2000; Djuretic et al., 2007; Kano et al., 2008). Most importantly, T-bet directly acts on the *Ifng* gene, as a transcriptional activator and by inducing changes of the chromatin structure on the *Ifng* locus (Chang and Aune, 2005; Hatton et al., 2006; Tong et al., 2005;

Mullen et al., 2001; Beima et al., 2006).

T-bet cooperates with several other factors in inducing a transcriptionally favorable chromatin structure on the *Ifng* locus (Fig. 1.2C). The transcription factor Ets1 binds to the *Ifng* promoter and *Ets1* deficient cells are defective in IFN-γ re-expression, but not in primary IFN-γ production (Grenningloh et al., 2005). Another transcription factor, "H2.0-like homeobox" (Hlx) is induced by T-bet and cooperates with T-bet in a feed-forward loop (Mullen et al., 2002; Martins et al., 2005). Hlx is required for T-bet dependent Th1 differentiation and seems to inhibit expression of the IL-4 receptor (Mullen et al., 2002; Mikhalkevich et al., 2006). Another transcription factor that has been reported to cooperate with T-bet is "runt-related transcription factor 3" (Runx3) (Djuretic et al., 2007). Like Ets1, Runx3 is dispensable for T-bet dependent induction of primary IFN-γ, but enhances IFN-γ re-expression (Djuretic et al., 2007). Runx3 is induced by T-bet and therefore Runx3 and T-bet also form a feed-forward loop, similar to the T-bet/Hlx interaction. T-bet and Runx3 do not only induce IFN-γ, but they also cooperate to repress IL-4 by binding to the *Il4* silencer (Djuretic et al., 2007; Naoe et al., 2007).

Epigenetic mechanisms in Th1 differentiation

The differentiation of naïve Th-cells towards an effector phenotype is accompanied by stable changes of the chromatin structure of lineage-specific cytokine genes (Wei et al., 2009). During Th1 differentiation, the *Ifng* locus acquires a permissive chromatin structure, while during Th2 differentiation inhibitory histone modifications accumulate and the DNA at the *Ifng* promoter is CpG-methylated (Jones and Chen, 2006; Chang and Aune, 2005; Chang et al., 2007). Decreased histone-acetylation has been described in cells lacking either T-bet or STAT4 (Avni et al., 2002; Fields et al., 2002; Chang and Aune, 2005) and ectopic expression of T-bet results in increased histone acetylation at the *Ifng* locus (Shnyreva et al., 2004). Moreover, T-bet interacts with a H3K27-histone demethylase, which removes repressive histone marks, and with a H3K4-methyl-transferase, which establishes permissive histone modifications (Miller et al., 2008). In addition, T-bet can even bind to the CpG-methylated *Ifng* promoter and induce removal of a co-repressor complex that is recruited to methylated DNA (Tong et al., 2005). This ability might enable T-bet to induce IFN-γ expression even in Th2 cells. STAT4 contributes to opening of the chromatin structure on the *Ifng* locus by recruiting a chromatin remodeling complex (Zhang and Boothby, 2006). During Th2 differentiation, on the contrary, STAT6 (activated by IL-4) and GATA3 bind to the *Ifng* locus to induce repressive histone modifications and silence the gene (Chang and Aune, 2007).

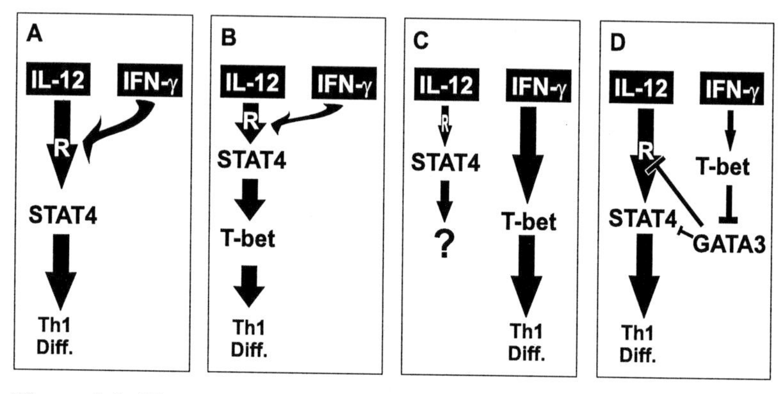

Figure 1.3. How IL-12 and IFN-γ might induce Th1 differentiation. (**A**) In the 1990ies, before the discovery of T-bet, it was proposed that STAT4, activated by IL-12, induces Th1 differentiation. IFN-γ would then be required to upregulate the IL-12 receptor. (**B**) When T-bet was discovered as a potent Th1 inducing transcription factor, IL-12 was proposed to drive Th1 differentiation through its putative ability to induce T-bet expression. (**C**) Later it was shown that IFN-γ, but not IL-12 could induce T-bet. This finding then raised the question of why IL-12 signaling was required for Th1 differentiation. (**D**) To solve this discrepancy, it was suggested that STAT4 was the main Th1 inducing transcription factor and that T-bet was only required to inhibit GATA3, which would otherwise repress the IL-12 receptor and STAT4.

In the previous sections, the regulatory interactions between genes that control Th1 differentiation were summarized, such as cytokines, their receptors and transcription factors. These known control mechanisms were used in this study as a starting point to develop a mathematical model of the gene-regulatory network driving Th1 differentiation. This model describes the regulation of Th1 specific genes during primary activation of naïve T-cells, but the ultimate goal was to understand, how Th1 cells are stabilized. This question is addressed in chapter 6. Previously proposed models of Th1 differentiation that attempt to distinguish the roles played by IL-12 and IFN-γ are described in the next section.

Models of Th1 differentiation

Already in the early 1990s it was discovered that IL-12 drives Th1 differentiation and that IFN-γ is required for IL-12 dependent priming, but cannot drive differentiation in the absence of IL-12 (Scott, 1991; Seder et al., 1993; Macatonia et al., 1993; Schmitt et al., 1994). Since IL-12 activates STAT4, the latter was proposed to be the central transcription factor driving differentiation. Analysis of *Stat4* deficient T-cells supported this idea (Thierfelder et al., 1996; Kaplan et al., 1996). Based on these observations, it was proposed that IL-12 would activate STAT4, which would then induce Th1 differentiation (scheme in Fig. 1.3A). IFN-γ was thought to be required to induce expression of IL-12Rβ2, which would then allow STAT4 activation by IL-12 (Szabo et al., 1997; Rogge et al., 1997). In this model, STAT4 activation and therefore also Th1 differentiation are prevented in the presence of IL-4, because it represses IL-12Rβ2. To test this model, cells that constitutively express IL-12Rβ2 were analyzed (Heath et al., 2000; Nishikomori et al., 2000). However, also in these cells Th1 differentiation could not be induced by IL-12. Therefore, it was concluded that STAT4 is insufficient to drive Th1 differentiation.

In 2000, the transcription factor T-bet was found to drive Th1 differentiation even in the presence of Th2 differentiation signals and in the absence of IL-12 (Szabo et al., 2000). It was hypothesized that IL-12/STAT4 would drive Th1 differentiation though induction of T-bet, which would then imprint the Th1 phenotype (Fig. 1.3B) (Glimcher and Murphy, 2000). However, later it was found that IL-12 dependent induction of T-bet expression was mediated by IFN-γ, independently of IL-12 (Afkarian et al., 2002; Lighvani et al., 2001). It still remained puzzling, why IL-12 is required for differentiation, if IFN-γ alone is able to induce the Th1 master transcription factor T-bet (Fig. 1.3C).

To explain the importance of IL-12 and STAT4, an alternative model was proposed by Strober and colleagues (Usui et al., 2003, 2006). In that model, STAT4, activated by IL-12, controls imprinting of the *Ifng* gene. The main function of T-bet is the downregulation of GATA3, which otherwise would inhibit Th1 differentiation by repression of STAT4 and IL-12Rβ2 (Fig. 1.3 D). However, T-bet has been shown to be required for Th1 differentiation even when induction of GATA3 is prevented by blocking IL-4 signaling (Szabo et al., 2002; Zhang and Boothby, 2006). Therefore, the mechanisms that control stable Th1 differentiation remain incompletely understood and the relative contributions of IL-12 and IFN-γ in that process are still unclear.

1.4 Mathematical modeling of Th-cell differentiation

Although gene-regulatory networks have been modeled more extensively for simpler organisms (Tomlin and Axelrod, 2007; Hasty et al., 2001), some attempts have been made to develop mathematical models of Th-cell differentiation. A mathematical model of Th2 differentiation showed that IL-4 upregulates GATA3, which then maintains its own expression through a positive feedback loop (Höfer et al., 2002). This model demonstrated that GATA3 auto-activation could be the molecular basis of Th2 phenotype stability in the absence of the differentiation signal IL-4. A later extension of this model, the so-called symmetric model, assumed a similar mechanism for the stabilization of Th1 cells through a positive feedback loop of T-bet (Mariani et al., 2004; Yates et al., 2004; Mendoza, 2006). Although, for some time T-bet had been assumed to directly induce its own expression (Mullen et al., 2001), this positive feedback regulation was later shown to be mediated by IFN-γ (Lighvani et al., 2001; Afkarian et al., 2002). It is unlikely that stabilization of the Th1 phenotype relies on an autocrine feedback loop, which would be rather sensitive to the immediate surrounding of the cell (as shown for IL-2 signaling in Busse, 2009). Alternatively, a so-called asymmetric model was proposed, where T-bet is constitutively expressed, when not repressed by GATA3 (Mariani et al., 2004). However, there is no experimental support for the hypothesis that T-bet is upregulated as soon as GATA3 is repressed. Thus, most existing models remain speculative, because they have not been validated experimentally.

1.5 Objectives of this thesis

Scientific aim

The overall aim of this work was to elucidate the structure of the gene-regulatory network driving Th1 differentiation and to understand, how this network controls stabilization of the Th1 phenotype. As described in the previous sections, the cytokines, receptors and transcription factors involved in Th1 differentiation mutually regulate each others activity in a rather complex manner. Although many details of the so formed regulatory network have been investigated, central questions remained unsolved:

- How is expression of the Th1 master transcription factor controlled?
- Which signals are critical in the regulation of IL-12Rβ2 during Th1 differentiation?
- Why is IL-12 a much more potent Th1 inducing signal than IFN-γ, although IFN-γ controls expression of the Th1 master transcription factor T-bet?

Approach and experimental design

To dissect the functioning of the Th1-differentiation network, experiments were closely combined with mathematical modeling. For the experimental measurements, an *in vitro* differentiation system was chosen that allowed tight control over the signals received by the cells. Naïve Th-cells were isolated from mice and differentiated in cell culture in the presence of IL-12 and a TCR-stimulus. Although the use of antigen-presenting cells would be a more physiological stimulus, in the present study TCR activating antibodies (αCD3) were used that allow a tight control of the cytokine milieu present in the culture. To observe the Th1 regulatory network in isolation, the alternative Th2 pathway was blocked in all experiments by addition of IL-4 specific blocking antibodies to the culture. In this culture system, mRNA expression kinetics of key Th1 specific genes were measured simultaneously over the entire time course of Th1 differentiation.

In parallel, a mathematical model of the Th1 regulatory network was developed, using ordinary differential equations. Because this work aimed at defining the central processes of Th1 differentiation, it was attempted to identify the simplest model that could account for the experimental observations. An initial literature-based model described regulation of only three genes that were known to play a central role: *Ifng*, *Il12Rb2* and *Tbx21* (Murphy and Reiner, 2002). The mathematical model was used to simulate expression kinetics, which were then compared to the experimental data. Disagreement between simulation and experiment suggested that important parts of the biological network were missing in the model. Through an iterative process, of hypothesizing new interactions, testing them experimentally and incorporating them in the mathematical description, a completed model of the Th1 differentiation network was developed. To validate the model specific regulatory interactions were interrupted experimentally with blocking antibodies, pharmacological inhibitors, or knock-out mice, and the resulting changes in expression kinetics were compared to simulations.

In a last step, it was analyzed, how the gene-regulatory network controlled differentiation. Stabilization of the Th1 phenotype was measured as the ability to re-express IFN-γ upon a secondary TCR-stimulus in the absence of IL-12, after six days of primary activation. Statistical tools were then used to link gene regulation during the differentiation process to phenotype stabilization.

Chapter 2

Literature-based model of Th1 differentiation

In the first part of this chapter, a mathematical model of the Th1 differentiation network is developed based on the available literature. In the second part, experimental measurements of the expression kinetics of Th1 specific genes are presented and compared to model simulations.

2.1 Model description

The model was expected to explain the transcriptional regulation of central Th1 specific genes in response to IL-12, autocrine IFN-γ and TCR stimulation over the time course of differentiation. The regulatory networks that mediate the effects of IL-12 and IFN-γ have been described in the introduction and are again summarized in Fig. 2.1A+B. The events involved in signal transduction (gray) were not modeled explicitly, because they proceed with much faster kinetics than transcriptional regulation (see Appendix A.1+A.2). IL-12 was usually added to the culture and was therefore assumed to be available at constant levels, while **IFN-γ** expression was explicitly modeled, because it acts in an autocrine manner. Transcriptional repression of IFN-γR was omitted, because it occurs rapidly after onset of stimulation, resulting in a constant low expression level throughout the time course of differentiation (see Appendix A.1). Transcriptional regulation of **IL-12Rβ2**, however, was known to occur on the time scale of days (Ouyang et al., 1998; Yamane et al., 2000) and was therefore one of the three genes included in the model. The third player in the literature-based model is **T-bet**. Other transcription factors acting downstream of T-bet, were not included (Fig. 2.1C), because they seem not to be required for primary IFN-γ expression, but only for memory expression, which is addressed in chapter 6. Signals and genes that control other lineages were also omitted, since in all measurements that

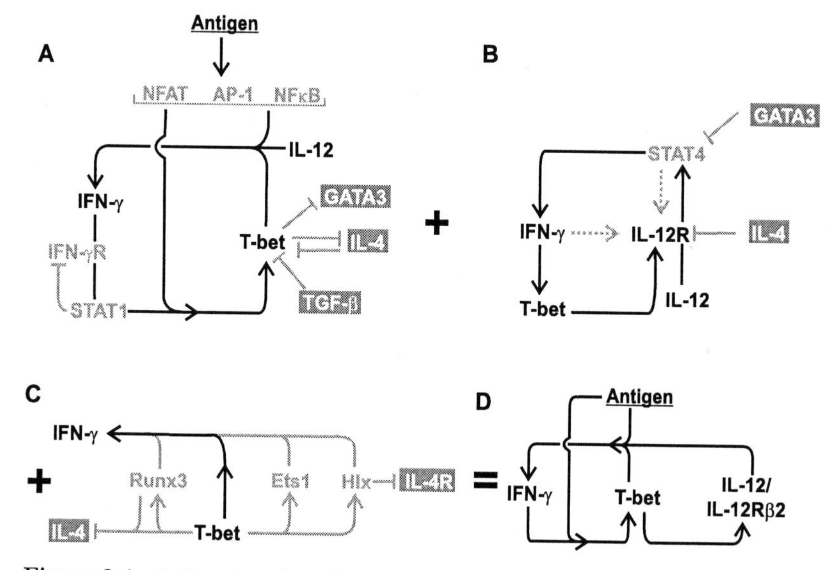

Figure 2.1. A literature based network model of the regulation of T-bet, IFN-γ and the IL-12Rβ2 during primary Th1 activation. (A-C) Regulatory networks of IFN-γ signaling, IL-12 signaling and T-bet, described in more detail in Fig. 1.2. Gray colored factors and regulatory interactions are not explicitly included in the model. **(D)** Literature based model of the Th1 differentiation network: IL-12 induces expression of IFN-γ in synergy with antigenic signals and T-bet, where the signal strength downstream of IL-12 is determined by the expression level of the IL-12Rβ2 subunit. Then, IFN-γ acting in an autocrine manner can upregulate T-bet expression in synergy with antigen signaling. Therefore, T-bet and IFN-γ form an autocrine positive feedback loop. T-bet also upregulates the IL-12Rβ2 subunit.

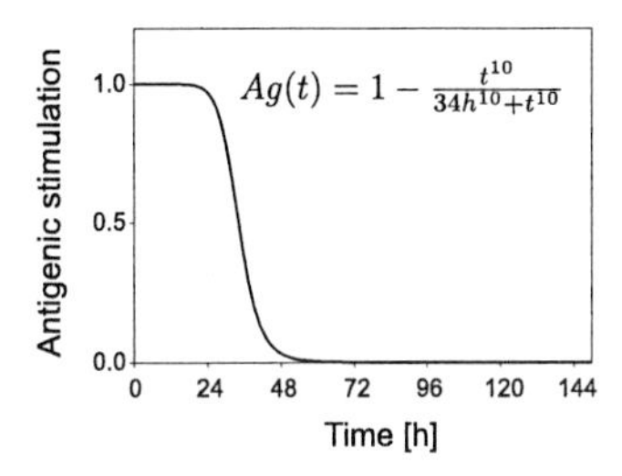

Figure 2.2. The kinetics of antigenic stimulation. The average relative strength of the antigen stimulation that was assumed in all simulations is plotted over time.

were used for development of the model differentiation of alternative fates was blocked (2.1A+B, gray). The network structure emerging after the described simplifications is shown in Fig. 2.1D.

To understand the gene expression dynamics that could be produced by this network, a mathematical description was developed in the form of a system of ordinary differential equations. All models used in this part are summarized in Table A.2 in Appendix A.2. The model describes production and degradation of protein and mRNA of T-bet, IFN-γ and IL-12Rβ2 (Rec). Translation and degradation rates were assumed proportional to mRNA and protein concentrations, respectively. Translation was assumed to proceed with a constant rate β, identical for all genes. Therefore protein levels were modeled as:

$$\frac{\mathrm{d}Protein}{\mathrm{d}t} = \beta \cdot mRNA - \delta \cdot Protein \tag{2.1}$$

where δ denotes the protein degradation rate constant. To describe transcriptional regulation, the concentration of transcription factors activated downstream of external signals was assumed to be proportional to the signal strength. Moreover, it was assumed that binding of the transcription factors to DNA equilibrated rapidly, and that transcription rates were proportional to transcription factor binding. Then, signal-dependent transcription rates can be described as saturating, hyperbolic (Michaelis-Menten-like) functions of the inducing signals (detailed derivation is given in the Appendix A.2). Independent activation by multiple signals was described by additive terms, while synergistic interaction was modeled by multiplicative terms.

Since naïve Th-cells express low levels of T-bet, but neither IL-12Rβ2 nor IFN-γ, only for T-bet a basal transcription rate was assumed (α_1). T-bet has been shown to indirectly promote its own expression through induction of IFN-γ, such that T-bet and IFN-γ form a positive feedback loop (Afkarian et al., 2002; Lighvani et al., 2001). As this feedback loop is only active in the presence of antigenic stimulation (Ag) (Lighvani et al., 2001), T-bet mRNA transcription was modeled as

$$\frac{d\,Tbet_{\mathrm{mRNA}}}{dt} = \alpha_1 + \alpha_2 \cdot \frac{Ag(t)}{K_1 + Ag(t)} \cdot \frac{IFN_{\mathrm{Prot}}}{K_2 + IFN_{\mathrm{Prot}}} - \gamma_{\mathrm{Tbet}} \cdot Tbet_{\mathrm{mRNA}} \quad (2.2)$$

IFN-γ expression in naïve Th-cells requires the joint action of IL-12, T-bet and antigenic signals. As external IL-12 concentrations in all experiments were saturating, IL-12 dependent signaling was controlled by the expression level of IL-12Rβ2 (Rec):

$$\frac{d\,IFN_{\mathrm{mRNA}}}{dt} = \alpha_5 \cdot \frac{Tbet_{\mathrm{Prot}}}{K_5 + Tbet_{\mathrm{Prot}}} \cdot \frac{Rec_{\mathrm{Prot}}}{K_6 + Rec_{\mathrm{Prot}}} \cdot \frac{Ag(t)}{K_7 + Ag(t)} - \gamma_{\mathrm{IFN}} \cdot IFN_{\mathrm{mRNA}} \quad (2.3)$$

Finally, IL-12Rβ2 is induced by T-bet:

$$\frac{d\,Rec_{\mathrm{mRNA}}}{dt} = \alpha_4 \cdot \frac{Tbet_{\mathrm{Prot}}}{K_8 + Tbet_{\mathrm{Prot}}} - \gamma_{\mathrm{Rec}} \cdot Rec_{\mathrm{mRNA}} \quad (2.4)$$

In the used experiments and *in vivo*, antigenic stimulation (Ag) is transient and is usually present for around 48 hrs. Consequently, the antigen stimulus was described by a time-dependent input function Ag(t) that dropped to 0 after 48 hrs of stimulation (Fig. 2.2).

2.2 Experimental testing of the literature-based model

To test, whether the literature-based model was able to explain the regulation of T-bet, IFN-γ and IL-12Rβ2, mRNA kinetics of these genes were measured over a full time course of Th1 polarization *in vitro*. Naïve murine $CD4^+$ T-cells were stimulated for 48 hrs with antibodies to CD3 and CD28, mimicking TCR stimulation by cognate antigen. To induce Th1 polarization, IL-12 and blocking antibodies to IL-4 were added to the culture. The expression kinetics of T-bet, IFN-γ and IL-12Rβ2 were quantified by RT-PCR (Fig. 2.3, circles). T-bet mRNA showed biphasic expression kinetics with two maxima at 24 hrs and 120 hrs after the onset of stimulation. IL-12Rβ2 was upregulated slowly with expression levels remaining low in the first two days. IFN-γ mRNA levels remained rather low in the first 24 hrs and showed a single, sharp peak after 48 hrs that dropped back to low signals after 72-96 hrs.

The model parameters were fitted simultaneously to the mRNA kinetics of T-bet, IL-12Rβ2 and IFN-γ (for details of the fitting procedure see Section 4.2 and Section 13.1). The literature-based model was unable to explain the experimentally observed kinetics (Fig. 2.3). In particular it failed to

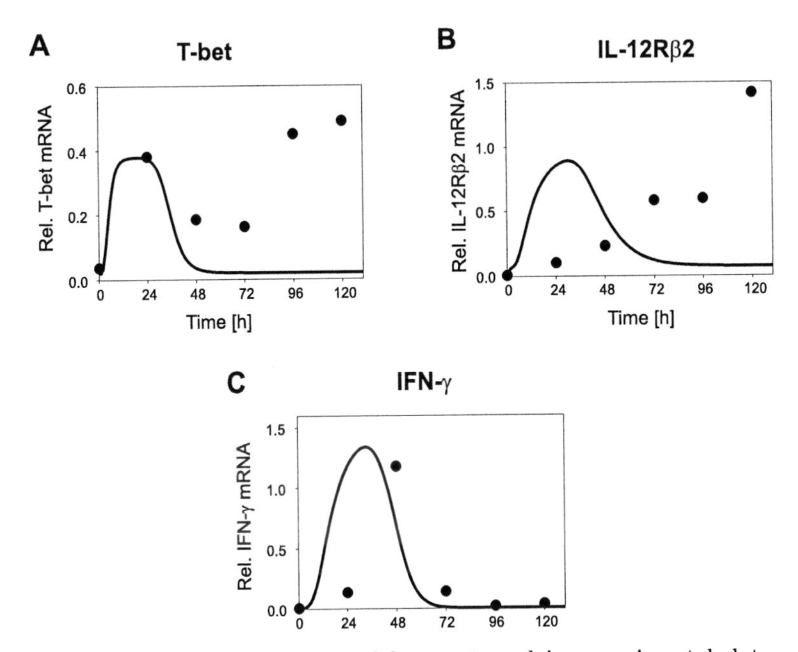

Figure 2.3. The one-loop model cannot explain experimental data. circles: Expression of T-bet (A), IL-12Rβ2 (B) and IFN-γ (C) mRNA, measured over five days of Th1 differentiation using naïve Th cells, isolated from C57BL/6 mice. A representative example of five independent experiments is shown. lines: Simulation (best fit) based on the one-loop (scheme in Fig. 2.1).

account for three central features: (1) The second T-bet peak after 120 hrs, (2) the slow induction of IL-12Rβ2 and (3) the low expression level of IFN-γ in the first 24 hrs. The model failure can be easily understood: Simultaneous stimulation with IFN-γ and antigenic signals was required for T-bet upregulation in the model. As antigenic stimulation was only present in the first 48 hrs, T-bet expression in the late phase of priming could not be explained. IL-12Rβ2 expression was induced by T-bet and followed therefore closely the kinetics of T-bet in the simulation. Consequently, a single IL-12Rβ2 peak occured in the simulation instead of the experimentally observed slow increase in the expression level. In the case of IFN-γ, a single peak was observed in experiment and simulation, but the exact timing was be reproduced by the model.

In summary, T-bet, IL-12Rβ2 and IFN-γ are expressed with distinct kinetics during Th1 differentiation. The literature-based model was unable to account for several features of the expression kinetics, such as two-peaked T-bet expression, slow induction of IL-12Rβ2 and low IFN-γ expression in the first 24 hrs. Therefore the literature-based model was concluded to be an incomplete description of the the regulation of T-bet, IL-12Rβ2 and IFN-γ, which was therefore analyzed in detail experimentally. The results are presented in the following chapters.

Chapter 3

Experimental analysis of the regulation of T-bet, IL-12Rβ2 and IFN-γ

3.1 T-bet regulation

As the literature-based model was unable to explain the observed biphasic expression of T-bet (Fig. 2.3A), the mechanisms controlling T-bet expression were examined experimentally. First, the role of cytokine signals was addressed and second the role of antigenic stimulation.

Regulation by IFN-γ and IL-12

Under standard Th1 inducing conditions (added IL-12, and autocrine IFN-γ signaling), T-bet mRNA showed two maxima at 24 hrs and 120 hrs after the onset of stimulation (Fig. 3.1A, solid line). To define the role of IFN-γ in T-bet regulation, IFN-γ signaling was interrupted using cells lacking the IFN-γ receptor 1 chain (*Ifngr*$^{-/-}$, Fig. 3.1B, solid line). This perturbation eliminated the first peak of T-bet expression, corroborating the previous observation that T-bet expression at early time points (24-48 hrs) is IFN-γ dependent (Afkarian et al., 2002; Lighvani et al., 2001). However, the second T-bet peak was still present, albeit the expression was reduced. To understand, which signals controlled T-bet expression in the late phase, *Ifngr* deficient cells were activated in the absence of IL-12 and the second T-bet peak disappeared (Fig. 3.1B, dotted line). These data showed that the early expression of T-bet during Th1 priming ($<$48 hrs) was driven by IFN-γ, whereas in the late phase ($>$72 hrs) T-bet expression directly depended on IL-12. Thus T-bet expression was examined in the absence of IL-12 in wild-type cells, with addition of recombinant IFN-γ to avoid potential indirect

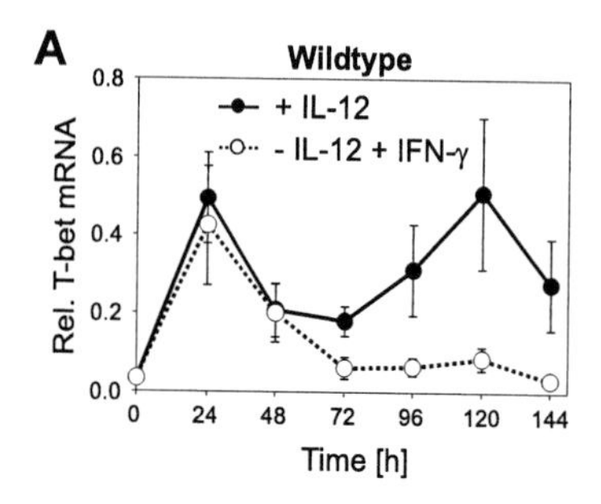

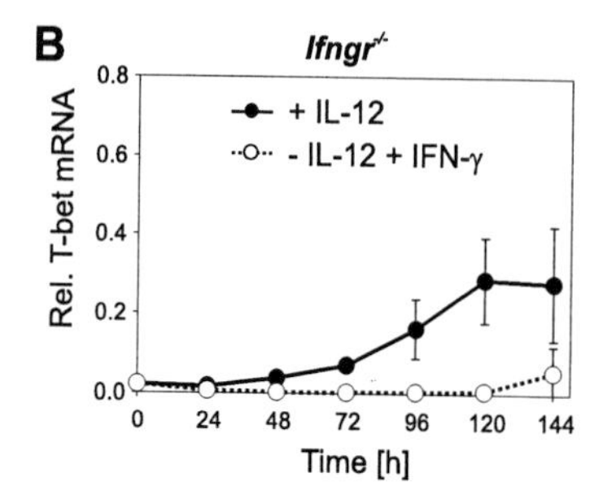

Figure 3.1. T-bet expression in the first 48 hrs of primary activation requires IFN-γ signaling, whereas later expression is directly induced by IL-12. Naïve Th cells from C57BL/6 **(A)** or *Ifngr*$^{-/-}$ mice on a C57BL/6 background **(B)** were stimulated with platebound CD3-specific antibodies for 48 hrs, and antibodies to CD28 and IL-4 were included in the medium. The cells were activated in the presence of IL-12 (solid lines) or IFN-γ and blocking antibodies to IL-12 (dotted lines). T-bet mRNA was quantified over six days using quantitative RT-PCR. Mean and s.d. of five independent experiments are shown.

effects due to reduced IFN-γ production when IL-12 was missing. The first T-bet peak remained unchanged, but the second peak was strongly reduced to a low basal expression (Fig. 3.1A, dotted line). It was concluded that IL-12 induces T-bet expression independent of IFN-γ, specifically in the late phase of priming.

Molecular mechanism of IL-12 dependent T-bet expression

To understand, which transcription factor might be responsible for IL-12 dependent T-bet expression, the role of STAT4 was investigated. For CD8$^+$ T-cells it has been reported that STAT4, the main transcription factor activated by IL-12, binds to an enhancer element on the *Tbx21* locus, 12 kilobases upstream of the transcriptional start site (Yang et al., 2007). Chromatin immunoprecipitation (ChIP) was used to test, whether STAT4 would also bind to this enhancer element in CD4$^+$ cells. Indeed, STAT4 bound to the *Tbx21* enhancer, but strong binding was only observed in the late phase of priming, when T-bet expression is IL-12 dependent (Fig. 3.2A, black bars). Moreover, STAT4 binding required the presence of IL-12 (Fig. 3.2A, compare black and gray bars) and no binding was observed with control primers that detect binding to a region 3 kb upstream of the transcription start site (Fig. 3.2B). This result suggests that IL-12 directly induces T-bet through activation of STAT4.

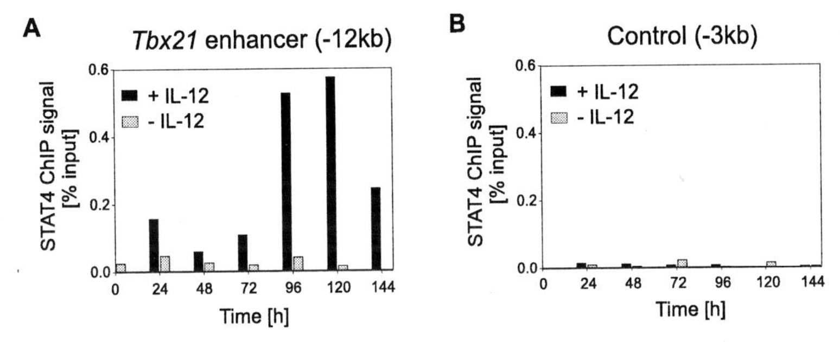

Figure 3.2. STAT4 binds to the *Tbx21* enhancer in the late phase of priming. Cells were treated as in Fig. 3.1 and chromatin was isolated for ChIP assay and precipitated with STAT4-specific antibodies. Quantitative PCR was performed using primers specific for the *Tbx21* enhancer element (**A**) and for a control sequence 3 kb upstream of the transcriptional start site (**B**). Bound STAT4 is expressed as percentage of the input used for the ChIP assay. The results are representative of two independent experiments.

Taken together, T-bet is expressed in two distinct waves during Th1 priming. For the first T-bet wave, IFN-γ signaling is required but IL-12 is dispensable, whereas the second wave is driven predominantly by an IL-12-dependent and IFN-γ-independent T-bet inducing mechanism and coincides with binding of STAT4 to the *Tbx21* enhancer.

Regulation of T-bet by antigenic signals

In the previous sections, IFN-γ was identified as an early (<48 hrs) and IL-12 as a late (>72 hrs) inducer of T-bet expression. Next, the mechanisms that restricted the activity of these signals to distinct phases were investigated. It was hypothesized that the antigen signal may be involved in the temporal control of T-bet expression, because it is terminated after 48 hrs of stimulation. Therefore the role of TCR signals in IFN-γ dependent T-bet expression was analyzed in detail. The measurements were taken after five hrs of stimulation because prolonged exposure to IFN-γ in the absence of TCR-signals induced cell death. In naïve Th-cells a low basal T-bet expression level was observed that remained unchanged, when the cells received a TCR-stimulus (αCD3) in the absence of IFN-γ signaling (Fig. 3.3A). Stimulation of naïve Th-cells in the presence of IFN-γ, by contrast, resulted in a ~20-fold induction of T-bet (Fig. 3.3A). This induction was decreased ~2-3-fold, when the cells were stimulated with IFN-γ alone (Fig. 3.3A). To show that the decrease in T-bet mRNA in the absence of antigenic signals was significant,

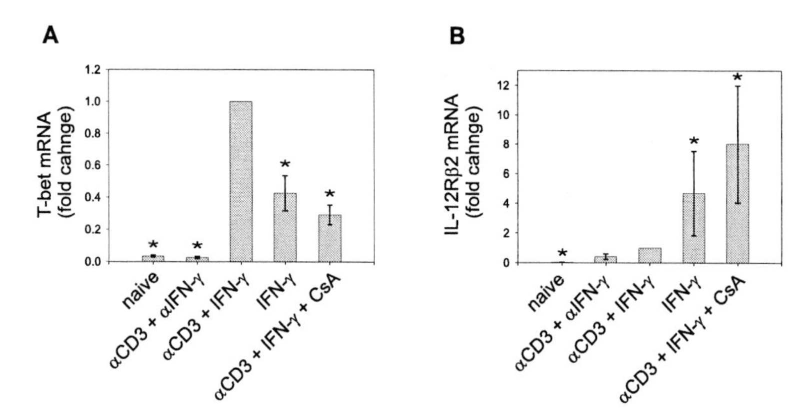

Figure 3.3. Antigenic signals induce T-bet, but repress IL-12Rβ2 expression. Naïve Th-cells from Balb/c mice were stimulated for five hrs with a TCR stimulus and IFN-γ (αCD3+IFN-γ). Expression levels were compared to conditions where either IFN-γ signaling was blocked with specific antibodies (αCD3+αIFN-γ) or where the TCR stimulus was removed, either by culturing the cells in absence of αCD3 (IFN-γ), or by addition of the calcineurin inhibitor Cyclosporine A (50 nM) that partially inhibits signaling downstream of the TCR (αCD3+IFN-γ+CsA). T-bet (A) and IL-12βR2 (B) mRNA was quantified using RT-PCR and the fold change in expression relative to stimulation with CD3-specific antibodies and IFN-γ was calculated. Mean and s.d. of four-six experiments are shown. *The fold change was significant in a paired t-test, $p<0.01$.

a paired t-test was applied to the logarithm of the expression levels. These results show that T-bet can be induced by IFN-γ signaling alone, but expression levels are increased further, when the TCR stimulus is also present. Therefore, T-bet is induced by the synergistic action of IFN-γ and TCR-signals, in agreement with previous studies (Lighvani et al., 2001). Thus the termination of antigen signaling appears to limit the initial, IFN-γ-driven wave of T-bet expression. This conclusion is also supported by the fact that even the addition of recombinant IFN-γ to the culture did not prolong IFN-γ dependent T-bet expression beyond 48 hrs (Fig. 3.1A, dotted line).

In the next step, it was analyzed, which TCR-dependent signaling pathway might be involved in IFN-γ dependent T-bet induction. To test, whether NFAT transcription factors mediated the TCR-effect, the pharmacological inhibitor cyclosporine A (CsA), which inhibits the NFAT-activating phosphatase calcineurin, was used. Addition of CsA to the culture stimulated with IFN-γ and αCD3 resulted in a ~3-fold reduced induction of T-bet, similarly to the stimulation in the absence of antigenic signals (Fig. 3.3A). This observation suggests that NFAT transcription factors are involved in T-bet induction.

The presented results show that T-bet expression in the early phase of priming is induced by the synergistic action of IFN-γ and the antigen/NFAT pathway. However, it remained unclear why IL-12 dependent T-bet expression was restricted to the late phase of primary activation. Since the measurements of IL-12Rβ2 mRNA (Fig. 2.3) and STAT4-binding to the *Tbx21* enhancer (Fig. 3.2A) revealed that both are strongly increased in the late phase of activation (>48 hrs), low STAT4 activity might limit IL-12 dependent T-bet expression in the early phase of activation.

3.2 IL-12Rβ2 regulation

Antigen-dependent repression of IL-12Rβ2

The IL-12-dependent mode of T-bet expression was not observed in the early phase of activation (Fig. 3.1B, solid line). This might result from the low expression level of the IL-12 receptor β2 chain in the first days of priming (Fig. 2.3, circles, middle panel). This hypothesis is also supported by previous reports showing that the expression level of IL-12Rβ2 limits downstream signaling of IL-12 and STAT4 activation (Szabo et al., 1997; Afkarian et al., 2002). Since the antigen signal is limited to the early phase, it might be responsible for repression of IL-12Rβ2.

To test this hypothesis, IL-12Rβ2 expression was compared in the presence and absence of antigenic signals after five hrs of stimulation (see above). A basal induction of IL-12Rβ2 was observed when stimulating naïve Th-cells with CD3-specific antibodies (Fig. 3.3B). Addition of IFN-γ did not significantly increase the expression level (Fig. 3.3B), although it has been reported to enhance IL-12Rβ2 expression through induction of T-bet (Afkarian et al., 2002). Surprisingly, IFN-γ without CD3 stimulation resulted in ~4-fold higher IL-12Rβ2 mRNA expression, suggesting the repression of the *Il12rb2* gene by TCR signaling (Fig. 3.3B). This increase was shown to be statistically significant by applying a paired t-test to the logarithms of the expression levels ($p<0.01$).

To investigate the signaling pathways involved in antigen-dependent repression of IL-12Rβ2, again the calcineurin inhibitor CsA was used. CD3 stimulation with simultaneous application of CsA resulted in a ~8-fold increase of IL-12Rβ2 mRNA expression (Fig. 3.3B). This observation suggests that TCR-dependent IL-12Rβ2 repression is mediated by the calcineurin/NFAT pathway. It has been reported previously for human T-cells that NFAT1 binds to the *Il12rb2* promoter (van Rietschoten et al., 2001). However, NFAT1 binding could not be detected in any of the candidate binding sites identified by sequence analysis (Appendix A.2).

In summary, the slow induction of IL-12Rβ2 during primary Th1 activation can be attributed to a repressive effect of antigenic signaling that is mediated by NFAT through a not yet identified mechanism. This initial repression of IL-12Rβ2 expression can also explain why IL-12 dependent expression of T-bet was only observed in the late phase of priming.

Induction of IL-12Rβ2 by T-bet

After the antigen signal was identified as a negative regulator of IL-12Rβ2 expression, its positive regulators were investigated in the next step. Previous studies on the regulation of IL-12Rβ2 had yielded partially contradictory results, showing that under certain conditions, IL-12Rβ2 can be induced by IFN-γ, IL-12 or T-bet (Szabo et al., 1997; Smeltz et al., 2002; Rogge et al., 1997; Chang et al., 1999). To distinguish the roles played by IL-12 and IFN-γ, the IL-12Rβ2 expression kinetics were measured in the presence of IL-12, IFN-γ, or of both signals. When both IFN-γ and IL-12 were present in the culture, IL-12Rβ2 mRNA was induced with slow kinetics and reached high levels in the late phase of priming (Fig. 3.4A, black solid line). In the absence of IL-12, expression levels were strongly reduced (gray solid line), while *Ifngr* deficient cells showed only a small defect in IL-12Rβ2 induction (black dotted line). When both, IL-12 and IFN-γ signaling were absent, no induction of IL-12Rβ2 was detected. These observations show that IL-12 can upregulate expression of its own receptor efficiently, while IFN-γ can only induce rather low expression levels.

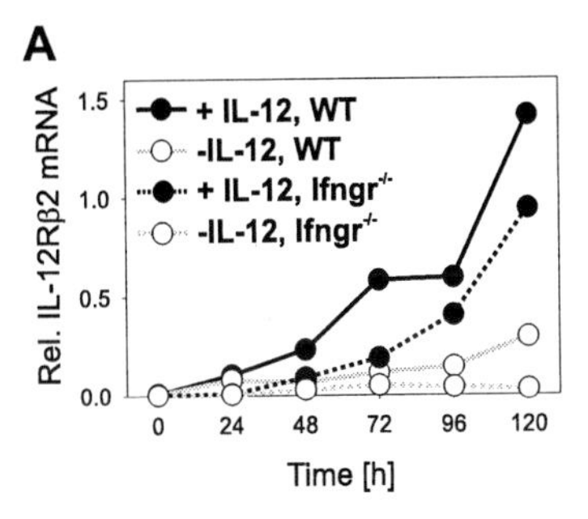

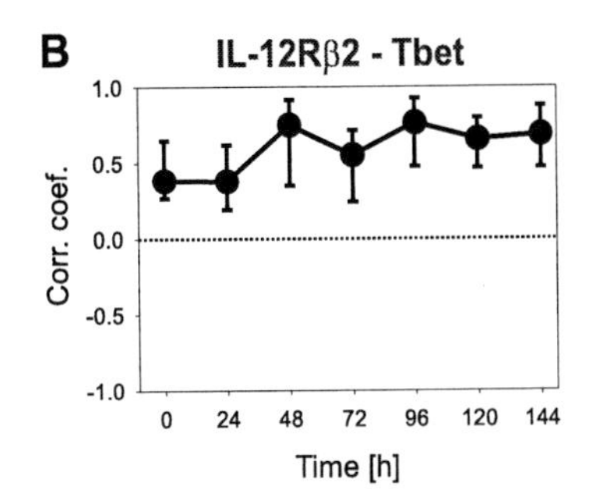

Figure 3.4. Control of IL-12Rβ2 by T-bet. (A) Naïve Th-cells, isolated from C57BL/6 (solid lines) or *Ifngr*$^{-/-}$ mice (dotted lines), were primed in the presence (black) or absence of IL-12 (gray). IL-12Rβ2 mRNA was quantified using RT-PCR. The data shown is representative of five independent experiments. **(B)** The experiment described in (A) was performed five times and T-bet as well as IL-12Rβ2 mRNA kinetics were assessed. For each time point, the correlation coefficient between the expression levels of these two genes was calculated (20 data points). The error bars indicate 95 % confidence intervals, calculated with a bootstrapping approach (see section 13.2).

If one considers that T-bet can induce IL-12Rβ2 expression even in the absence of IL-12 and IFN-γ signals (Afkarian et al., 2002), the observed IL-12 dependent induction of IL-12Rβ2 might well be mediated by T-bet, which is induced by IL-12, specifically in the late phase of priming (section 3.1). In addition, even the low IL-12Rβ2 levels induced by IFN-γ alone could be attributed to T-bet, which is also present at low levels under these conditions (compare Fig. 3.1 and 3.4A). If the effects of IL-12 and IFN-γ were mediated by T-bet, a correlation between the expression levels of IL-12Rβ2 and T-bet should be observed. To test this hypothesis, the expression kinetics of these two genes were quantified in five independent experiments. At all time points a significant positive correlation was observed between the expression levels of T-bet and IL-12Rβ2 (Fig. 3.4B). This robust and strong correlation supports the view of T-bet as the main inducer of IL-12Rβ2.

In summary, IL-12 as well as IFN-γ can contribute to IL-12Rβ2 induction through their ability to induce T-bet expression. But as IL-12Rβ2 is repressed in the early phase of priming, IL-12, the main inducer of T-bet in the late phase, is also the main regulator of IL-12Rβ2 expression.

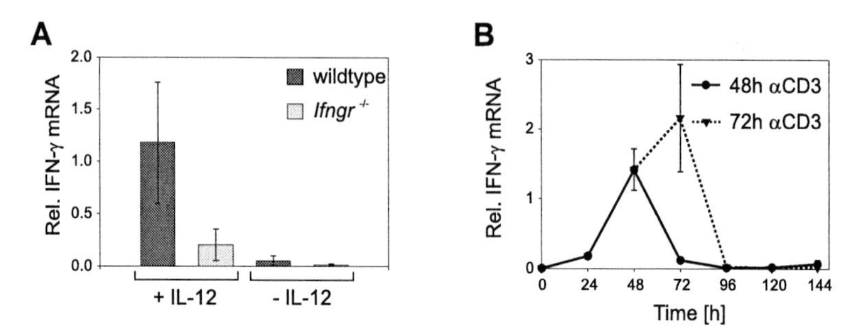

Figure 3.5. IFN-γ expression requires the joint action of IL-12, T-bet and TCR-signals. (A) Naïve Th-cells, isolated from C57BL/6 (dark bars) or *Ifngr*$^{-/-}$ mice (light bars), were stimulated for 48 hrs with IL-12 (+IL-12) or IFN-γ and blocking antibodies to IL-12 (-IL-12). IFN-γ mRNA was quantified using quantitative RT-PCR. Mean and s.d. of five independent experiments are shown. **(B)** Naïve Th-cells from C57BL/6 mice were cultivated on plate-bound antibodies to CD3 for 48 hrs (solid lines) or for 72 hrs (dotted lines). IFN-γ mRNA was quantified using quantitative RT-PCR. Mean and s.d. of three independent experiments are shown.

3.3 IFN-γ regulation

In the experimental measurements of mRNA kinetics during Th1 differentiation presented above, IFN-γ shows a single, rather sharp peak after 48 hrs (Fig. 2.3). The simulation using the literature-based model also showed a single peak, but the low mRNA levels in the first 24 hrs that were observed in the experiment could not be recapitulated in the simulation (Fig. 2.3C). To understand which signals were required for induction of IFN-γ expression, IFN-γ mRNA was quantified after 48 hrs of stimulation (when the maximal level is reached), in wildtype and *Ifngr*$^{-/-}$ cells, in the presence and absence of IL-12 (Fig. 3.5A). IFN-γ expression was activated in a synergistic fashion by IL-12 and IFN-γ signaling. In particular, the much reduced IFN-γ mRNA expression measured in *Ifngr* deficient cells is likely to be due to the lack of T-bet induction (see Fig. 3.1B), which is required for IFN-γ production (Szabo et al., 2002). These results suggest that high level IFN-γ expression requires the joint action of IL-12 and IFN-γ/T-bet.

Next, the temporal control of IFN-γ expression was addressed. Why was high-level IFN-γ expression restricted to the time window between 24 and 72 hrs, although IL-12 and IFN-γ/T-bet were present throughout the culture period? As shown in the previous sections, IL-12 signaling is low in first 48 hrs, because IL-12Rβ2 is repressed by antigenic signals (see 3.2). Consequently, the low expression level of IFN-γ during the first day can be attributed to reduced IL-12 signaling. The termination of IFN-γ expression after 48 hrs might be due to removal of the antigen stimulus at that time. To test this hypothesis, the duration of antigenic stimulation was varied and the resulting changes in IFN-γ mRNA kinetics were analyzed. Prolonging TCR-stimulation by an additional 24 hrs extended the period of IFN-γ expression by the same time interval, demonstrating its strict dependence on antigen signaling (Fig. 3.5B). Therefore, it was concluded that high IFN-γ expression in naïve Th-cells requires the joint action of IL-12 and TCR-signaling as well as T-bet. Consequently, high IFN-γ expression was limited to a short time window around 48 hrs of primary activation, before TCR signaling was terminated and after IL-12Rβ2 expression had reached sufficient levels.

In summary, the detailed investigation of how expression of T-bet, IL-12Rβ2 and IFN-γ expression are regulated during Th1 priming yielded the following results:

1. T-bet expression is induced by the synergistic action of IFN-γ and TCR signaling in the early phase of priming. In the late phase, IL-12, mediated by STAT4, directly maintains T-bet expression independent of IFN-γ.

2. IL-12Rβ2 expression is repressed by TCR-signaling in the early phase, mediated by NFAT. The main positive regulator of IL-12Rβ2 is T-bet, mediating IL-12 and IFN-γ dependent effects.

3. High-level IFN-γ expression is only observed, when IL-12, IFN-γ/T-bet and TCR-signals are present simultaneously.

Chapter 4

The two-loop model

In the previous chapters, a number of interactions were confirmed that had already been included in the literature-based model. In addition, two hitherto unknown mechanisms were described that seem to be crucial elements of the regulatory network controlling Th1 differentiation, namely the induction of T-bet through the IL-12/STAT4 pathway, and antigen-dependent repression of the IL-12Rβ2 subunit. Fig. 4.1 schematically shows the network structure that is formed, when these two regulatory links are added to the literature-based model. In this completed network, a second positive feedback loop controls T-bet expression: T-bet enhances expression of the IL-12 receptor and thereby augments IL-12 signaling, which is in turn required to maintain T-bet expression. This network was termed the two-loop model, in contrast to the literature-based one-loop model, where T-bet expression is controlled by a single feedback loop with IFN-γ. In the two-loop model, T-bet expression is controlled by two positive feedback loops, mediated by IFN-γ and IL-12 signaling, respectively. The antigen stimulus acts as a switch between the two pathways, because it is required for IFN-γ dependent T-bet

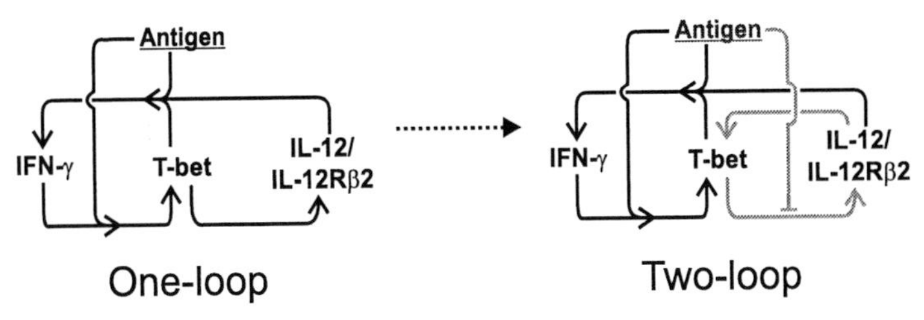

Figure 4.1. The two-loop model. Schematic representation of the literature-based one-loop model and the completed two-loop model, where the newly found interactions, such as IL-12-dependent T-bet expression and antigen-dependent repression of IL-12Rβ2, were integrated.

expression, but inhibits IL-12 signaling.

4.1 The two T-bet controlling feedback loops

To test, whether the IFN-γ and IL-12 mediated feedback loops were indeed central to T-bet regulation, their interconnecting pathways were perturbed experimentally. If the positive feedback loop was indeed the central mechanism controlling T-bet and IFN-γ in the early phase of priming, then blocking one interconnecting pathway, e.g. IFN-γ signaling, should prevent expression of both genes. As predicted, the first wave of T-bet expression was completely abolished in the absence of IFN-γ signaling (Fig. 4.1A left, compare solid and dotted lines). Surprisingly, the second IL-12 dependent wave was also affected, since it was induced with a delay of ~24 hrs.

In the late phase, the IL-12R/T-bet feedback was proposed to be the central regulatory element controlling expression of these two genes. To test this hypothesis, cells were cultured in the absence of IL-12 and, as expected, the first T-bet wave was left intact, but the second wave was strongly diminished together with IL-12Rβ2 levels (Fig. 4.2A+B, compare dashed and solid lines). Under all three conditions, expression of IL-12Rβ2 was upregulated synchronously with the second, IL-12 dependent wave of T-bet expression, supporting the existence of the proposed T-bet/IL-12Rβ2 feedback loop (compare Fig. 4.2A+B). IL-12Rβ2 expression was strongly diminished in the absence of IL-12 signaling (Fig. 4.2B, dashed lines), while induction was delayed by ~24 hrs in *Ifngr* deficient cells (Fig. 4.2B, compare solid and dotted lines). Therefore IFN-γ seemed to accelerate the activation of the T-bet/IL-12R feedback loop. These results suggest that indeed early in differentiation a positive feedback loop with IFN-γ controls T-bet expression, while in the late phase the T-bet/IL-12Rβ2 plays a central role.

4.2 Two-loop model can explain experimental data

To understand, whether the two-loop model could also quantitatively account for the measured expression kinetics, two terms for the newly found interactions were added to the mathematical description of the one-loop model (Section 2.1). Equation 2.2, describing T-bet mRNA was completed by ad-

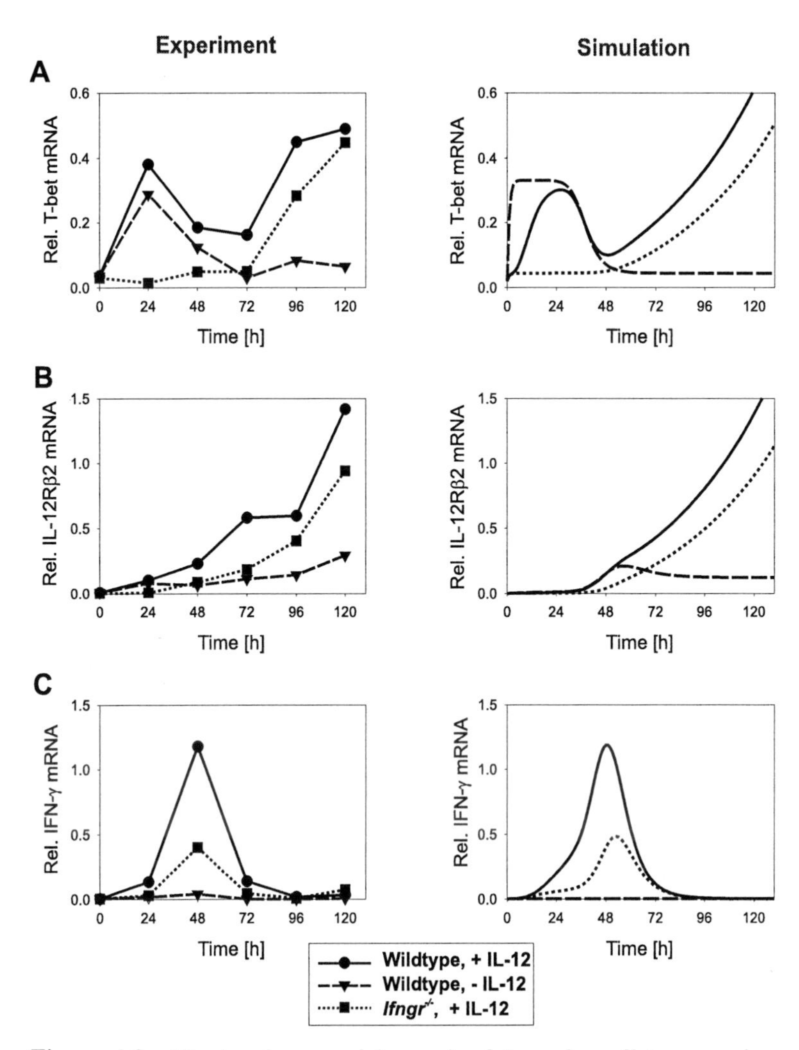

Figure 4.2. The two-loop model can simulate and predict expression kinetics under various experimental conditions. left: Naïve Th-cells, isolated from C57BL/6 (solid and dashed lines) or *Ifngr*$^{-/-}$ mice (dotted lines), were stimulated for five days with IL-12 (solid and dotted lines) or IFN-γ and blocking antibodies to IL-12 (dashed lines). T-bet (A), IL-12Rβ2 (B) and IFN-γ mRNA kinetics (C) were quantified. The experiment was performed three times, one representative example is shown. right: Expression kinetics of T-bet (A), IL-12Rβ2 (B) and IFN-γ mRNA (C) were simulated based on the two-loop model (Fig. 4.1) under Th1-inducing conditions (solid lines), and in the absence of IFN-γ (dotted lines) or IL-12 signaling (dashed lines). All parameters were simultaneously fitted to the experimental data.

dition of an IL-12R (Rec)-dependent term:

$$\begin{aligned}\frac{\mathrm{d}\,Tbet_{\mathrm{mRNA}}}{\mathrm{d}t} = & \quad \alpha_1 + \alpha_2 \cdot \frac{Ag(t)}{K_1 + Ag(t)} \cdot \frac{IFN_{\mathrm{Prot}}}{K_2 + IFN_{\mathrm{Prot}}} \\ & +\alpha_3 \cdot \frac{Rec_{\mathrm{Prot}}}{K_3 + Rec_{\mathrm{Prot}}} - \gamma_{\mathrm{Tbet}} \cdot Tbet_{\mathrm{mRNA}}\end{aligned} \quad (4.1)$$

To account for antigen-dependent repression of IL-12Rβ2, equation 2.4 of the one-loop model was modified as follows:

$$\frac{\mathrm{d}\,Rec_{\mathrm{mRNA}}}{\mathrm{d}t} = \alpha_4 \cdot \frac{Tbet_{\mathrm{Prot}}}{K_8 + Tbet_{\mathrm{Prot}}} \cdot \frac{K_4}{K_4 + Ag(t)} - \gamma_{\mathrm{Rec}} \cdot Rec_{\mathrm{mRNA}} \quad (4.2)$$

In the next step, the new model was fitted to the data. To constrain the kinetic parameters, mRNA and protein half-lives were set to reasonable values (all mRNA species: 40 min; T-bet and IL-12Rβ2 proteins: 7 hrs, Table 4.1). Since IFN-γ protein acts in an autocrine fashion in the model, a short effective half-life of 40 min was assumed because, upon secretion, IFN-γ leaves the vicinity of the cell rapidly by diffusion and through re-uptake by the cell. Using a simulated-annealing algorithm, all activation rate constants and half-saturation constants were fitted simultaneously to mRNA kinetics of all three genes in wildtype and *Ifngr*$^{-/-}$ ($\alpha_2 = 0$) cells, in the presence and absence ($\alpha_3 = \alpha_6 = 0$) of IL-12. Since mRNA levels were quantified relative to the house-keeping gene HPRT, all simulations and parameters use these relative units. Because the values found for K_3, the half-saturation constant of IL-12 dependent T-bet transcription, and K_8, the half-saturation constant of T-bet dependent transcription of IL-12Rβ2, were much higher than the values reached by Rec_{Prot} and $Tbet_{\mathrm{Prot}}$, respectively, it was possible to linearize the terms containing these parameters and to, thereby eliminate K_3 and K_8 from the model. Equations 4.1 and 4.2 were simplified to

$$\frac{\mathrm{d}\,Tbet_{\mathrm{mRNA}}}{\mathrm{d}t} = \alpha_1 + \alpha_2 \cdot \frac{Ag(t)}{K_1 + Ag(t)} \cdot \frac{IFN_{\mathrm{Prot}}}{K_2 + IFN_{\mathrm{Prot}}} \quad (4.3)$$

$$+\alpha_3 \cdot Rec_{\mathrm{Prot}} - \gamma_{\mathrm{Tbet}} \cdot Tbet_{\mathrm{mRNA}} \quad (4.4)$$

$$\frac{\mathrm{d}\,Rec_{\mathrm{mRNA}}}{\mathrm{d}t} = \alpha_4 \cdot Tbet_{\mathrm{Prot}} \cdot \frac{K_4}{K_4 + Ag(t)} - \gamma_{\mathrm{Rec}} \cdot Rec_{\mathrm{mRNA}} \quad (4.5)$$

The simplified model was again fitted to the experimental data (parameter values in Table 4.1) and could indeed quantitatively explain the expression kinetics of all three genes under Th1-inducing and under perturbed conditions (Fig. 4.2, compare left and right). To estimate confidence intervals for the parameters, a bootstrapping approach was used. New data sets were generated from the experimental data by addition of noise, estimated from

Table 4.1. Parameter values of the two-loop model

Activation rate constants	Parameter values	90% confidence interval
α_1	0.044 h^{-1}	[0.032 0.054]
α_2	0.42 h^{-1}	[0.29 1.28]
α_3	0.00051 h^{-1}	[0.00044 0.00060]
α_4	0.0028 h^{-1}	[0.0024 0.0035]
α_5	3.7 h^{-1}	[1.6 9.6]
Half-saturation constants		
K_1	0.46	[0.14 2.88]
K_2	2.1	[0.6 7.2]
K_4	0.013	[0.0073 0.020]
K_5	0.029	[$2.3 \cdot 10^{-8}$ 36]
K_6	66	[18 155]
K_7	0.014	[0.0061 0.025]
mRNA degradation		
γ_{Tbet}	1 h^{-1}	n.a.
γ_{Rec}	1 h^{-1}	n.a.
γ_{IFN}	1 h^{-1}	n.a.
Translation rate		
β	100 h^{-1}	n.a.
Protein degradation		
δ_{Tbet}	0.1 h^{-1}	n.a.
δ_{Rec}	0.1 h^{-1}	n.a.
δ_{IFN}	1 h^{-1}	n.a.

Table 4.2. Model Comparison

	Model A	Model B	Model C
s.s.r.$_{\text{experiment}}$	3.2	2.2	0.46
Δs.s.r.$_{\text{experiment}}$	2.96	1.92	0.22
Δs.s.r.$_{\text{bootstrap}}$	[0.002 0.011]	[-$7 \cdot 10^{-4}$ 0.12]	[-0.0018 0.0011]
p-value	$<$0.01	$<$0.01	$<$0.01

repeated measurements, and the best fit was determined for each data set. From the distribution of fitted parameter values, the 5th and 95th percentile was calculated to estimate 90 % confidence intervals (Table 4.1).

4.3 Model comparison

The two-loop model clearly performed better in explaining the experimental data than the one-loop model (compare Fig.s 2.3 and 4.2), suggesting that the completed model is a better representation of the biological network than the initial literature-based model. However, as the two-loop model is more complex in that it contains more parameters, its improved performance could also originate from a better fit of the measurement noise. To exclude this possibility a statistical model comparison was performed. In addition to the one-loop model (Model A, Fig. 4.3A), two other models were included in this comparison, each accounting for only one of the new features, such as TCR-dependent IL-12Rβ2 repression (Model B, Fig. 4.3B) and IL-12 dependent T-bet expression (Model C, Fig. 4.3C).

All four models (Models A-C and two-loop model, not simplified) were fitted to the experimental data. Using the sum of squared residuals (s.s.r.) as an estimator for the goodness of the fit, it was found that the two-loop model performed best (s.s.r.= 0.25, compare to s.s.r.$_{\text{experiment}}$ in Table 4.2). To quantify the improvement of the fit with the two-loop model compared to the other models Δs.s.r.$_{\text{experiment}}$ was calculated as the difference between the s.s.r. of model A/B/C and the two-loop model (Table 4.2). Then the hypotheses were tested that either Model A, B or C were the true models, while the two-loop model only fitted the data better because it contained more parameters (Timmer et al., 2004). Based on the best fits of models A, B and C, 100 new data sets for each model were generated in a bootstrap approach by adding measurement noise. Each of the bootstrap data sets was fitted with the model used to generate it and with the two-loop model, and Δs.s.r.$_{\text{bootstrap}}$ was calculated. In Table 4.2, minimal and maximal values

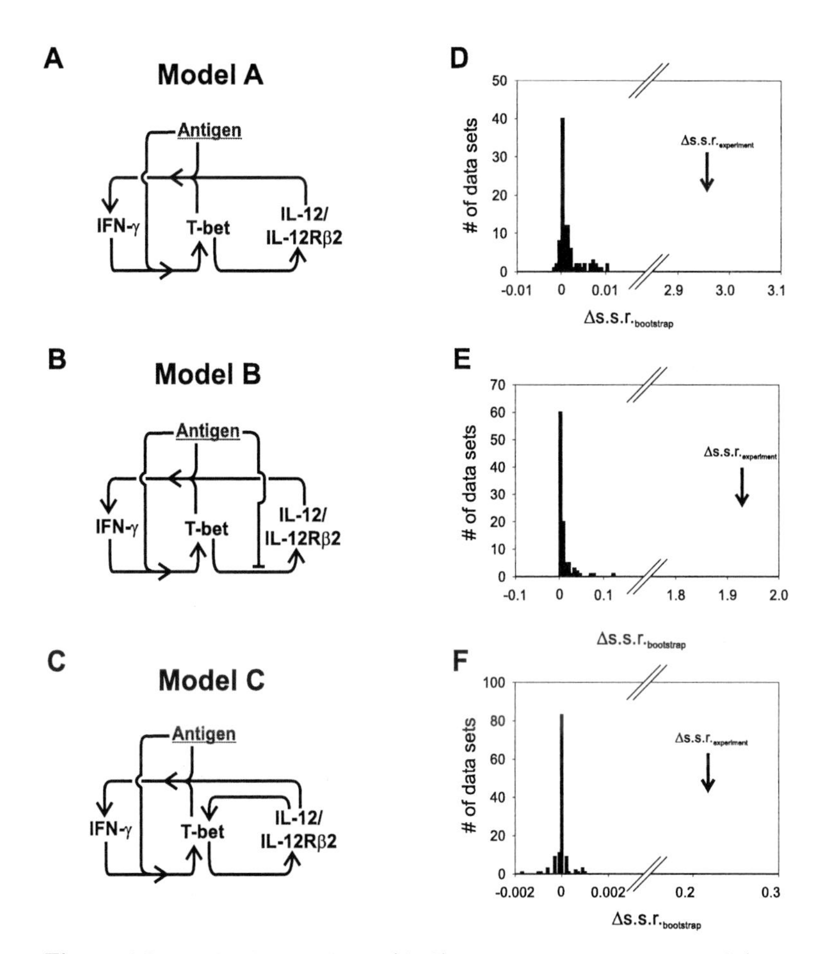

Figure 4.3. Model Comparison. (A-C) Network structure of Models A (one-loop model), B and C. **(D-F)** From each Model (A-C) 100 bootstrap data sets were generated. $\Delta s.s.r._{bootstrap}$ was calculated by subtraction of s.s.r. (sum of squared residuals) of the fits with the generating model and with the two-loop model. The distribution of $\Delta s.s.r._{bootstrap}$ for each model is shown. The arrow indicates $\Delta s.s.r._{experiment}$, which quantifies the improvement, when fitting the experimental data with the two-loop model instead of Model A, B or C, respectively.

of Δs.s.r.$_{\text{bootstrap}}$ are shown. For all three models (A,B,C) Δs.s.r.$_{\text{bootstrap}}$ of all 100 bootstrap data sets was much smaller than Δs.s.r.$_{\text{experiment}}$ (Fig. 4.3D-F). Therefore it was concluded with $p<0.01$ that Δs.s.r.$_{\text{experiment}}$ cannot be explained by the higher complexity of the two-loop model.

In summary, it was found that IL-12-induced T-bet expression and antigen dependent inhibition of IL-12Rβ2 are necessary to account for the expression kinetics of T-bet, IFN-γ, and IL-12Rβ2 during Th1 differentiation.

4.4 Model validation

Independent experiments

So far, the two-loop model has been developed based on a single data set. Next it was verified that the model would account for two more data sets from independent experiments. For another two *ex vivo* isolated batches of naïve Th-cells, T-bet, IL-12Rβ2 and IFN-γ kinetics were measured under Th1 conditions and with interrupting either IFN-γ signaling, IL-12 signaling, or both together (Fig. 4.4). While all major qualitative features of the expression kinetics, such as two-peaked T-bet expression with a single IFN-γ peak and delayed induction of IL-12Rβ2, were highly reproducible, the absolute expression levels differed substantially, in particular for IL-12Rβ2 (Fig. 4.4). To test, whether the two-loop model could account for all data sets, model parameters were fitted to each repetition. All three experiments could be fitted equally well with an s.s.r. of 0.24, 0.3 and 0.06 (Table A.1 in Appendix A.4).

Protein measurements

Up to this point, mRNA measurements have been used to develop the model. Three central features of the model were also verified on the protein level: IFN-γ dependence of early T-bet, IL-12 dependence of late T-bet, and slow induction of IL-12Rβ2. T-bet protein levels were measured by intracellular flow cytometry in wildtype and *Ifngr*$^{-/-}$ cells in the presence and absence of IL-12. In agreement with the mRNA measurements, T-bet expression in the early phase (40 hrs after onset stimulation) was only observed in wildtype cells, because it required IFN-γ signaling (Fig. 4.5A, black bars). By contrast, in the late phase of priming (at 120 hrs) T-bet was only detected in the presence of IL-12 (Fig. 4.5A, gray bars). These findings are in good agreement with model simulations (Fig. 4.5B) and support the notion that the two-loop model is a good representation of the gene-regulatory network active in Th-cells.

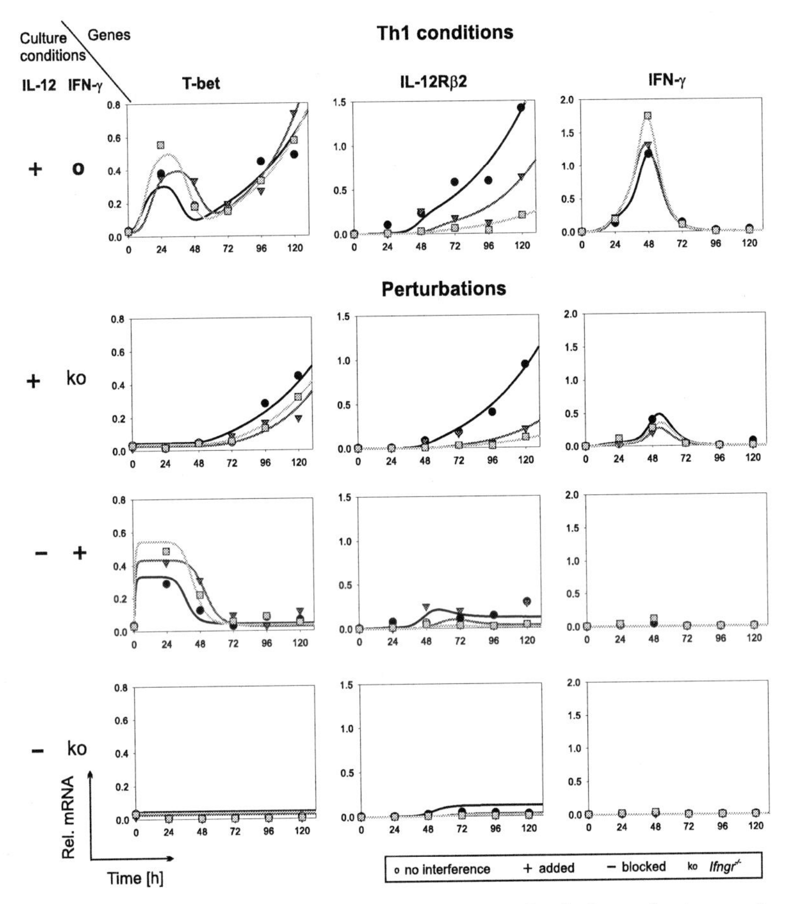

Figure 4.4. The two-loop model can account for independent experiments. Naïve Th-cells, isolated from C57BL/6 or *Ifngr*$^{-/-}$ mice, were stimulated for six days as indicated on the left. T-bet, IL-12Rβ2 and IFN-γ mRNA kinetics were quantified. The experiment was performed three times (triangles, squares, circles). The two-loop model was fitted to each experiment separately. The best fit for each data set is shown (lines).

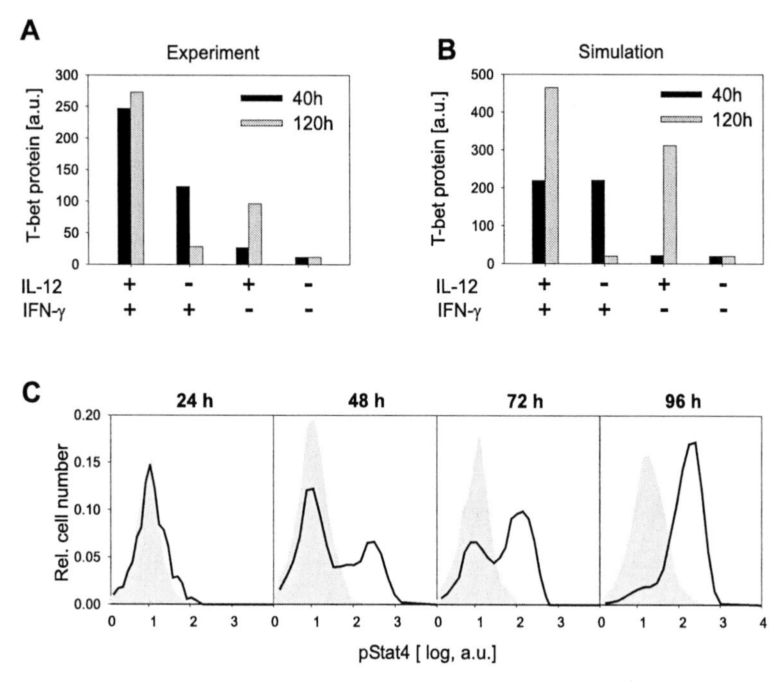

Figure 4.5. Model validation on the protein level. **(A)** Naïve Th-cells from wildtype C57BL/6 or *Ifngr*$^{-/-}$ mice were activated in the presence of IL-12 or IFN-γ (and blocking antibodies to IL-12). T-bet protein levels were quantified after 40 hrs and after 120 hrs using intracellular flow cytometry. The mean fluorescence signals were calculated and corrected for background staining using isotype control antibodies. The results shown are representative of two independent experiments. **(B)** The experiment described in (A) was simulated *in silico* using the two-loop model. **(C)** Naïve Th-cells from wildtype C57BL/6 mice were activated under Th1 inducing conditions. At the indicated time points, phosphorylated STAT4 was measured by intracellular flow cytometry (solid lines). To assess background staining, cells stimulated in the absence of IL-12 were used (shaded area). One representative example out of two independent experiments is shown. (a.u. arbitrary units)

In the next step, a central assumption in the model was verified experimentally, namely that IL-12Rβ2 mRNA levels would control IL-12Rβ2 protein expression on the cell surface, which in turn would determine the amount of activated STAT4 in the cell. If this was true, the kinetics of STAT4 activation should mirror the slow induction of IL-12Rβ2 mRNA (Fig. 3.4). To test this prediction, STAT4 phosphorylation was measured by intracellular flow cytometry between 24 and 96 hrs (Fig. 4.5C). No phosphorylation was detectable at 24 hrs; between 48 and 96 hrs the fraction of cells containing pSTAT4 increased continuously and STAT4 was phosphorylated in nearly all cells at the end of this period. Since the measured STAT4 activation kinetics resemble the induction of IL-12Rβ2 mRNA (compare to Fig. 3.4A), the extent of STAT4 phosphorylation seems indeed to be controlled by the number of IL-12Rβ2 transcripts.

Model predicts the response to antigen perturbation

In the two-loop model, the antigen stimulus directs the two waves of T-bet expression through activating the IFN-γ/T-bet loop, while inhibiting the IL-12/T-bet feedback. Can the model predict the response to a perturbation of TCR-stimulation? In simulations, inhibition of TCR signaling after 24 hrs resulted in accelerated induction of IL-12Rβ2 and advanced onset of the second wave of T-bet expression (Fig. 4.6A+B, left). Using CsA to block activation of the calcineurin-NFAT pathway, it was found that experimental inhibition of antigen-dependent signals after 24 hrs of TCR stimulation resulted precisely in the predicted effect (Fig. 4.6A+B, right). It was concluded that the antigen stimulus indeed acts as a switch to start up the T-bet/IL-12Rβ2 feedback loop.

In summary, the presented results suggest that the two-loop model describes the core network of Th1 differentiation. It can explain regulation of three central Th1-specific genes, namely *Tbx21*, *Ifng* and *Il12rb2*, under Th1 inducing conditions and with perturbation of the IL-12, IFN-γ and antigen signals. In contrast to the literature-based model, it captures all major features of the expression kinetics of these three genes.

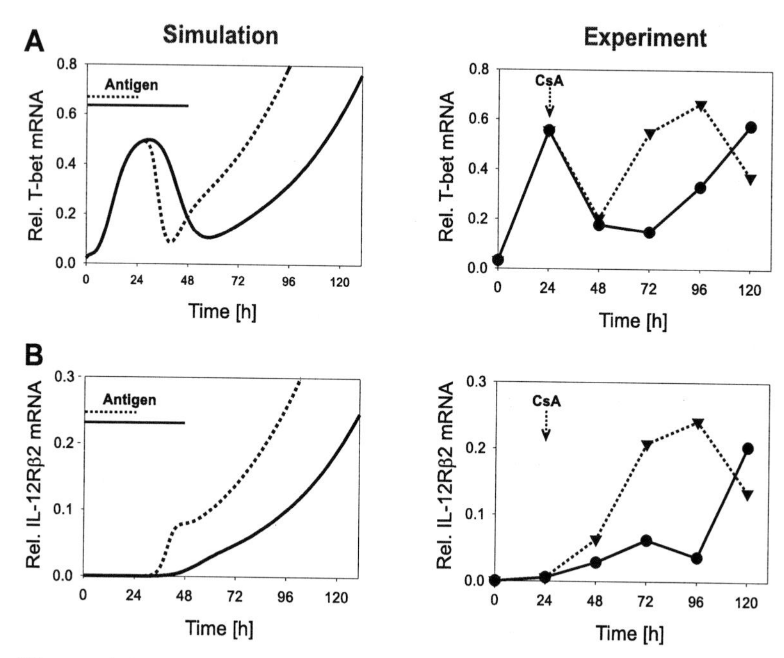

Figure 4.6. The duration of antigen signaling controls activation of the T-bet/IL-12Rβ2 feedback loop. left: Expression kinetics of T-bet (A) and IL-12Rβ2 mRNA (B) are compared *in silico* in cells stimulated with antigen for 24 hrs (dotted lines) or 48 hrs (solid lines) under Th1 inducing conditions. right: Naïve Th-cells, isolated from C57BL/6 mice, were cultured under Th1-inducing conditions and cyclosporine A (CsA) was added after 24 hrs of stimulation as indicated (triangles). T-bet (A) and IL-12Rβ2 mRNA (B) was quantified over five days of culture. One representative example out of three independent experiments is shown.

Chapter 5

Possible extensions of the two-loop model

In the previous chapter, a rather small, but highly convoluted gene-regulatory network was shown to be sufficient to explain the expression kinetics of T-bet, IFN-γ and IL-12Rβ2 under various culture conditions. Taking into account that a rather large number of genes have been implicated in Th1 differentiation (Wilson et al., 2009), it is somewhat surprising that such a small system of only three genes can account for so many experimental observations. In this section, possible extensions of the two-loop model are discussed, taking into account additional genes and alternative regulatory mechanisms.

5.1 GATA3-mediated STAT4 repression

The transcriptional regulation of IL-12Rβ2 has been discussed in some detail in the preceding sections. However, IL-12Rβ2 transcription is not the only level at which IL-12 signaling is tuned. Also expression (not only phosphorylation) of STAT4, the transcription factor activated by IL-12, is regulated in the context of Th-cell differentiation (Usui et al., 2003, 2006). STAT4 expression is repressed by the Th2-specific transcription factor GATA3 (Usui et al., 2003). This GATA3 dependent repression of STAT4 has been suggested to play a role also during differentiation of Th1 cells, where T-bet inhibits GATA3 (Usui et al., 2003, 2006). The model that was developed based on these observations was already discussed briefly in the introduction (Fig. 1.3D). According to this model, T-bet would upregulate STAT4 and IL-12Rβ2 indirectly, by inhibiting GATA3, which would otherwise repress these two genes (Fig. 5.1B). In this way T-bet would prevent down-regulation of STAT4 and IL-12Rβ2, which in turn would enhance IL-12 signaling and induce IFN-γ.

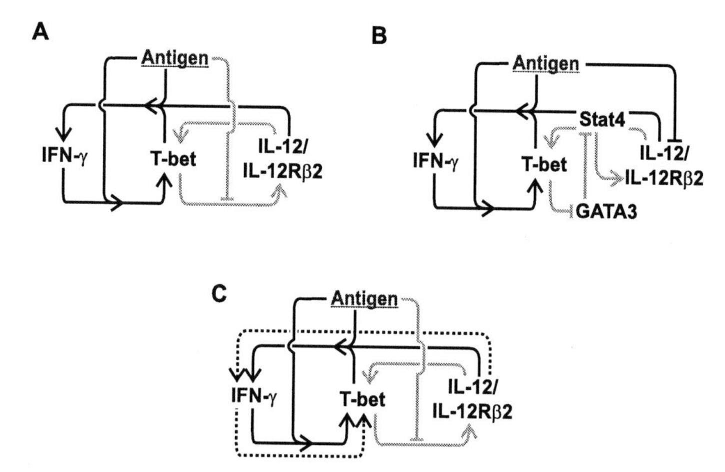

Figure 5.1. Extensions of the two-loop model. (A) Two-loop model. In model (B) T-bet acts only indirectly through inhibiting GATA3, which would otherwise repress STAT4 and IL-12Rβ2. In Model (C) antigen signaling is not strictly required for IFN-γ expression and IFN-γ dependent T-bet induction. Instead, low-level IFN-γ expression can be induced by IL-12 alone and IFN-γ can induce weak T-bet transcription independent of antigen signal (dotted lines).

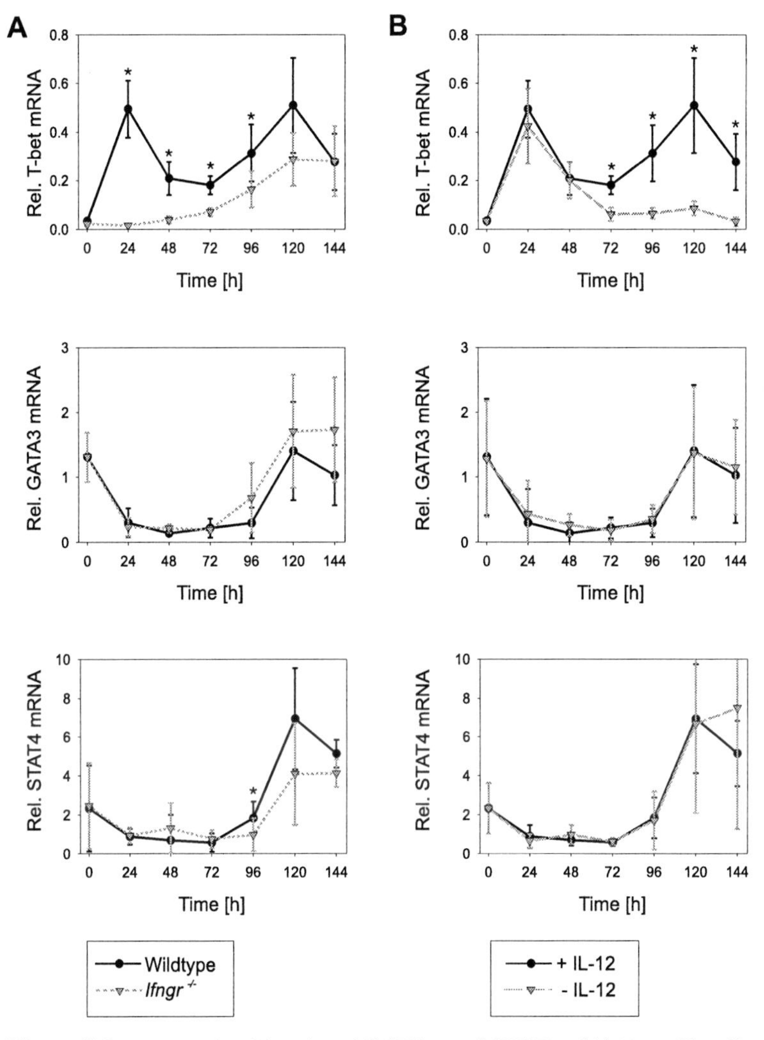

Figure 5.2. Expression kinetics of GATA3 and STAT4. (A) Naïve Th-cells from wildtype C57BL/6 (black) or *Ifngr*$^{-/-}$ mice (gray) were activated under Th1 inducing conditions. T-bet (top), GATA3 (middle) and STAT4 (bottom) mRNA was quantified over six days using RT-PCR. **(B)** Naïve Th-cells from C57BL/6 mice were activated in the presence of IL-12 (black) or IFN-γ and blocking antibodies to IL-12 (gray). T-bet (top), GATA3 (middle) and STAT4 (bottom) mRNA was quantified over six days using RT-PCR. Mean and s.d. of five independent experiments are shown. *In a paired t-test, the expression levels were significantly different between the two conditions analyzed, with $p<0.05$.

To test, whether induction of IFN-γ and IL-12Rβ2 by T-bet were independent of, or mediated by GATA3 (Fig. 5.1 compare A+B), STAT4 and GATA3 mRNA levels were measured in wildtype and *Ifngr* deficient cells in the presence and absence of IL-12 (Fig. 5.2). Both genes showed an initial drop of their expression levels upon stimulation that increased again in the late phase of priming. GATA3 only recovered the low levels present in naïve cells, while STAT4 was induced to levels, which were substantially above the initial expression. These temporal profiles were also observed in the absence of IFN-γ or IL-12 signaling. If T-bet function was indeed mediated by GATA3 and STAT4, high T-bet levels should result in low GATA3 expression and high STAT4 expression. However, in the experimental measurements this was not observed (Fig. 5.2). In particular, no difference in STAT4 and GATA3 mRNA expression was found, when comparing wildtype and *Ifngr* deficient cells in the first 48 hrs of stimulation, a phase when T-bet expression levels are vastly different (Fig. 5.2A). Similarly, at 72-120 hrs after the onset of stimulation, comparison of cells cultured in the presence and absence of IL-12, showed that, although the expression of T-bet differed greatly, there was no significant difference in the expression of STAT4 and GATA3 (Fig. 5.2B). Only when comparing *Ifngr* deficient cells with wildtype cells 96 hrs after the onset of stimulation in the presence of IL-12, a small but statistically significant decrease in STAT4 mRNA levels was observed, which could reflect GATA3 mediated inhibition of STAT4 expression. These data suggest that transcription of STAT4 and GATA3 is not controlled by IL-12, IFN-γ, or T-bet.

To substantiate our finding that T-bet, if at all, plays only a minor role in the regulation of STAT4 and GATA3, a correlation analysis was performed. As explained in more detail in section 3.2, data of five independent experiments, each including different culture conditions, was used to calculate the correlation coefficient between the expression levels of T-bet and STAT4 or GATA3 for each time point (Fig. A.4A in Appendix A.4). No significant correlation was found, not even a uniform trend towards a positive or negative correlation. To understand, whether the regulation of STAT4 and GATA3 was linked to other genes in the two-loop model, a similar correlation analysis was performed with IL-12Rβ2 and IFN-γ, but again no uniform trend was detected (Fig. A.4B+C in Appendix A.4). Surprisingly, a strong positive correlation, in particular during the first days of activation, was calculated between STAT4 and GATA3. This result stands in stark contrast to the proposed repression of STAT4 by GATA3. Therefore, the presented results suggest that during Th1 differentiation, when IL-4 signaling is blocked, either one of the genes induces expression of the other one, or, more likely, a common regulator controls expression of STAT4 and GATA3.

In summary, STAT4 and GATA3 are not regulated by the two-loop network during Th1 differentiation. Instead, their expression seems to be regulated by other pathways that might be related to antigenic stimulation or proliferation. Furthermore, regulation of IFN-γ and IL-12Rβ2 by T-bet is predominantly via the mechanisms of the two-loop model (Fig. 5.1A) and without a substantial effect via repression of GATA3 and STAT4 transcription.

5.2 IFN-γ dependent T-bet expression in the late phase

A major result of this project was the finding that T-bet expression in the late phase of activation requires IL-12 stimulation. However, as already mentioned in section 4.1, also IFN-γ signaling exerted a small effect on T-bet expression in the late phase. In wildtype cells activated in the absence of IL-12, the second T-bet wave is reduced to a low plateau, but expression levels are still significantly higher than in *Ifngr* deficient cells (Fig. 5.3A). This IFN-γ dependent plateau cannot be explained by the two-loop model. Maybe, IFN-γ can also induce T-bet in the late phase of priming, when antigenic stimulation has been terminated. This idea is also supported by the observation, described in section 3.1 that IFN-γ induces T-bet in the absence of TCR signaling, albeit to a lesser extent (Fig. 3.3A).

To understand, whether autocrine IFN-γ signaling might contribute to T-bet induction in the late phase, IFN-γ expression was analyzed during that period. In the absence of IL-12 signaling, IFN-γ mRNA levels were very low between 72 and 144 hrs after onset of stimulation and they were indistinguishable from the levels observed in naïve cells (Fig. 5.3B, green and blue). However, in the presence of IL-12, IFN-γ mRNA was significantly increased (for a statistical analysis see Fig. A.3 in Appendix A.4), although these levels were much lower than peak expression after 48 hrs (Fig. 5.3B, red and black). To test, whether the cells indeed produced IFN-γ protein in the late phase, expression was measured by intracellular flow cytometry. Under Th1 inducing conditions, the entire population expressed low IFN-γ levels, albeit variable quantities (Fig. 5.3C, solid line). In addition, a small fraction of cells showed a high expression level after 48 hrs, which resulted in the mRNA peak at that time point. When IL-12 and IFN-γ signaling were blocked (Th0), no IFN-γ expression was observed (Fig. 5.3C, dotted line). Therefore, IL-12 can induce low-level IFN-γ expression also in the absence of antigenic signals.

From the presented data it was concluded that the low levels of IFN-γ, expressed in the late phase upon IL-12 stimulation might increase T-bet levels

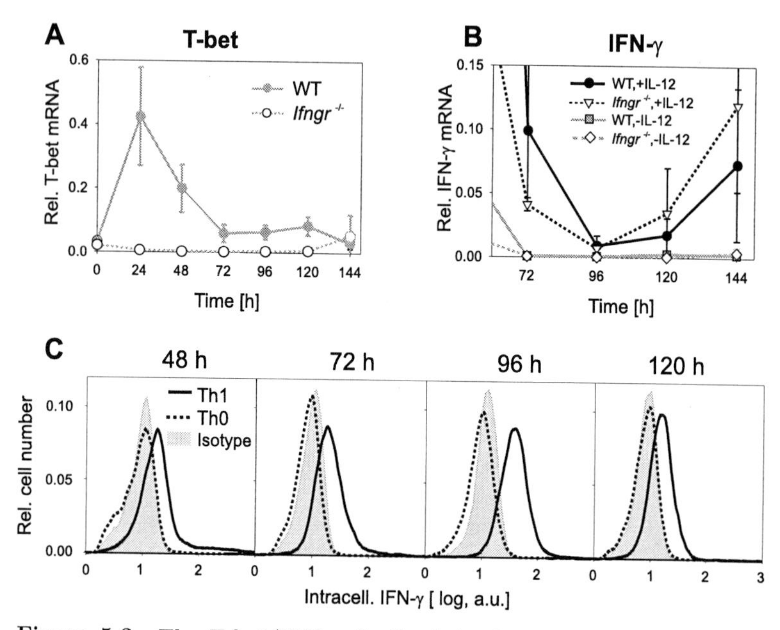

Figure 5.3. The T-bet/IFN-γ feedback in the late phase of priming. (A,B) Naïve Th-cells from C57BL/6 (solid lines) or *Ifngr*$^{-/-}$ mice (dotted lines) were stimulated in the presence of IL-12 (black) or IFN-γ and blocking antibodies to IL-12 (gray). T-bet (A) and IFN-γ (B) mRNA was quantified in five independent experiments, mean and s.d. are shown. (C) IFN-γ protein expression was measured with intracellular flow cytometry during Th1 differentiation. Cells cultured under Th1 inducing conditions (WT+IL-12, solid line) were compared with cells cultured in the absence of IL-12 and IFN-γ signaling (*Ifngr*$^{-/-}$, -IL-12, dotted line), and with isotype control staining (gray area).

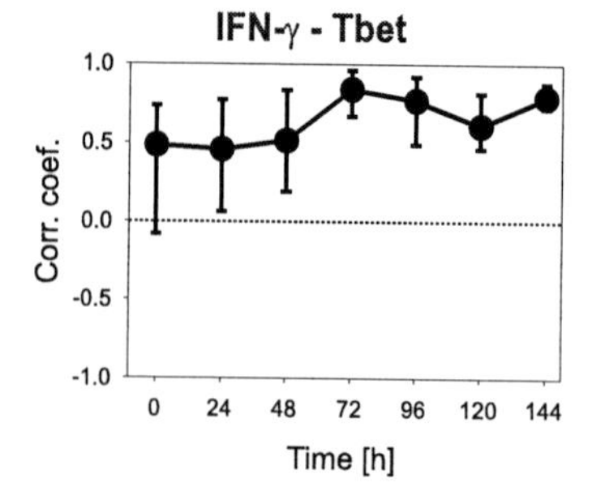

Figure 5.4. T-bet and IFN-γ levels correlate in the late phase of priming. As described in Fig. 3.4 the correlation coefficient between T-bet and IFN-γ levels was calculated for each time point, and 95% confidence intervals (error bars) were estimated through a bootstrapping approach.

through autocrine signaling. The proposed mechanism could be interpreted as an additional positive feedback loop: IL-12 induces IFN-γ, which induces T-bet, which in turn enhances IFN-γ expression through induction of the IL-12 receptor (Fig. 5.1 C). This additional feedback loop might be the reason, why IFN-γ increases T-bet expression in the late phase (Fig. 5.5A+B, left). IFN-γ dependent T-bet expression in the late phase was also observed in the absence of IL-12, because IFN-γ was added to the culture under these conditions. If indeed T-bet and IFN-γ would form a positive feedback loop also in the late phase of priming, a correlation between the expression of these two genes should be observed. This hypothesis was tested using five independent experiments and indeed a correlation coefficient between 0.6 and 0.9 was observed between 72 and 144 hrs (Fig. 5.4).

In the next step, the two-loop model was extended, so that it would account for IFN-γ dependent T-bet expression and IL-12 dependent IFN-γ expression also in the absence of antigenic signals (Fig. 5.1C). To describe antigen-independent T-bet induction by IFN-γ (with rate constant α_2'), equation 4.1 was modified as follows

$$\begin{aligned}\frac{\mathrm{d}\,Tbet_{\mathrm{mRNA}}}{\mathrm{d}t} &= \alpha_1 + \frac{IFN_{\mathrm{Prot}}}{K_2 + IFN_{\mathrm{Prot}}} \cdot \left(\alpha_2' + \alpha_2 \cdot \frac{Ag(t)}{K_1 + Ag(t)}\right) \\ &+ \alpha_3 \cdot \frac{Rec_{\mathrm{Prot}}}{K_3 + Rec_{\mathrm{Prot}}} - \gamma_{\mathrm{Tbet}} \cdot Tbet_{\mathrm{mRNA}}\end{aligned} \tag{5.1}$$

IL-12 dependent expression in the absence of antigenic stimulation (with rate constant α_5') was included in the equation describing IFN-γ mRNA regulation:

$$\begin{aligned}\frac{\mathrm{d}\,IFN_{\mathrm{mRNA}}}{\mathrm{d}t} &= \alpha_5 \cdot \frac{Rec_{\mathrm{Prot}}}{K_6 + Rec_{\mathrm{Prot}}} \cdot \left(\alpha_5' + \frac{Tbet_{\mathrm{Prot}}}{K_5 + Tbet_{\mathrm{Prot}}} \cdot \frac{Ag(t)}{K_7 + Ag(t)}\right) \\ &\quad - \gamma_{\mathrm{IFN}} \cdot IFN_{\mathrm{mRNA}}\end{aligned} \tag{5.2}$$

To test, whether the extended model could account for the T-bet plateau in the absence of IL-12 and IL-12 dependent IFN-γ expression in the late phase of differentiation, the extended model was fitted to the experimental data (Fig. 5.5). Indeed the extended model could reproduce the IFN-γ dependent increase in T-bet expression levels in the late phase in presence and absence of IL-12 (Fig. 5.5A+B). Moreover, it accounted for IL-12 dependent low-level IFN-γ expression in the late phase of priming (Fig. 5.5C). In summary, the model extension allowed the simulation of these two, so far not understood features of the measured kinetics and might therefore represent an important part of the regulatory network underlying Th1 differentiation.

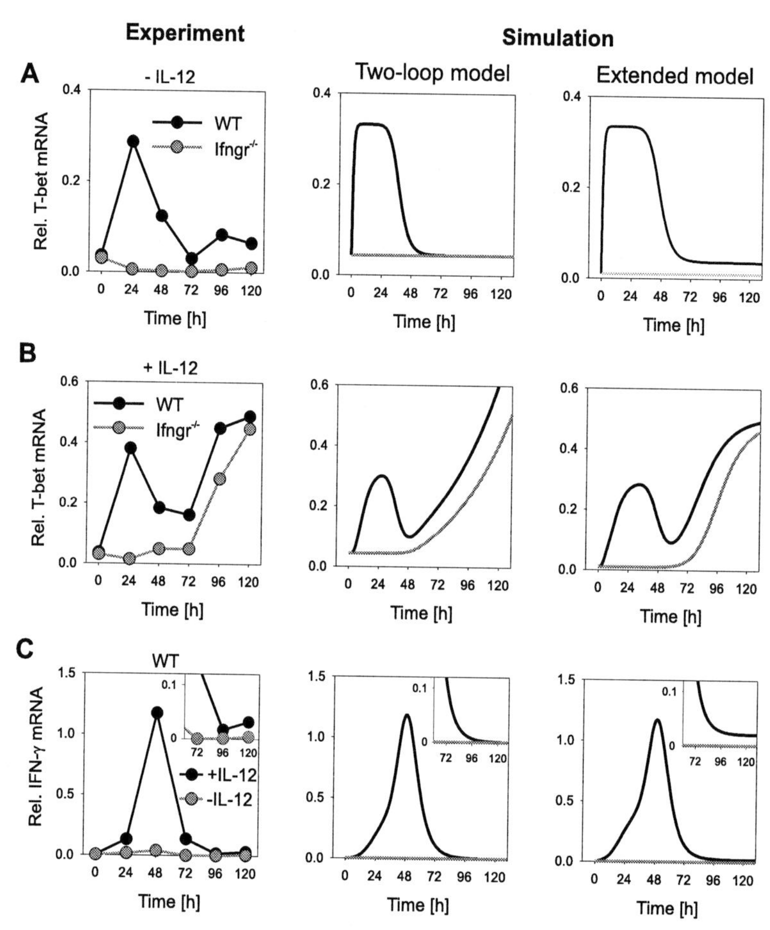

Figure 5.5. Extended model can explain IFN-γ dependent T-bet expression in the late phase. Experimental measurements (left column) are compared with simulations based on the two-loop model (middle) and based on the extended model (right), presented in Fig. 5.1C. **(A,B)** The extended model can account for increased T-bet levels in the late phase that are observed in wildtype cells compared to *Ifngr*$^{-/-}$ cells in the presence (B) and absence (A) of IL-12. **(C)** Low-level IL-12 dependent IFN-γ expression in the late phase of priming is reproduced by the extended model.

Chapter 6

Stabilization of the Th1 phenotype

In the previous chapters, the gene regulatory network active during the Th1 differentiation process was analyzed. In this last part, the question will be addressed of how this network controls the differentiated Th1 phenotype. Until now, the relative contributions of IL-12 and IFN-γ to Th1 differentiation remained enigmatic. IFN-γ controls expression of T-bet, but is not sufficient to drive Th1 differentiation, which requires IL-12 (Macatonia et al., 1993; Seder et al., 1993; Wenner et al., 1996). In chapter 3, it was shown that both, IFN-γ and IL-12 can induce T-bet expression, albeit during different phases of differentiation. So why is IL-12 a more potent Th1-polarizing signal than IFN-γ, although T-bet is induced by both cytokines?

6.1 Late T-bet is required IFN-γ for memory

IFN-γ controls T-bet expression in the early phase of priming, while IL-12 maintains expression in the late phase. Since IL-12 unlike IFN-γ is a strong Th1-inducing signal, the expression of T-bet during a specific time window might be a critical determinant for Th1 differentiation. To test this hypothesis, it was investigated how T-bet expression at different time points affected Th1 polarization. The polarization efficiency was quantified as the ability of the cells to produce the Th1-signature cytokine IFN-γ upon a second encounter of an antigen-mimicking stimulus (PMA+ionomycin). To test, whether T-bet expression during the first 24 hrs influenced polarization, cells that had expressed different amounts of T-bet in this initial phase were compared (Fig. 6.1A, black and dark grey lines). Through addition of recombinant IFN-γ, T-bet induction kinetics were accelerated, but no increase in the polarization efficiency was observed (Fig. 6.1B, compare black and dark grey). Next, T-bet expression was inhibited in the first 48 hrs using

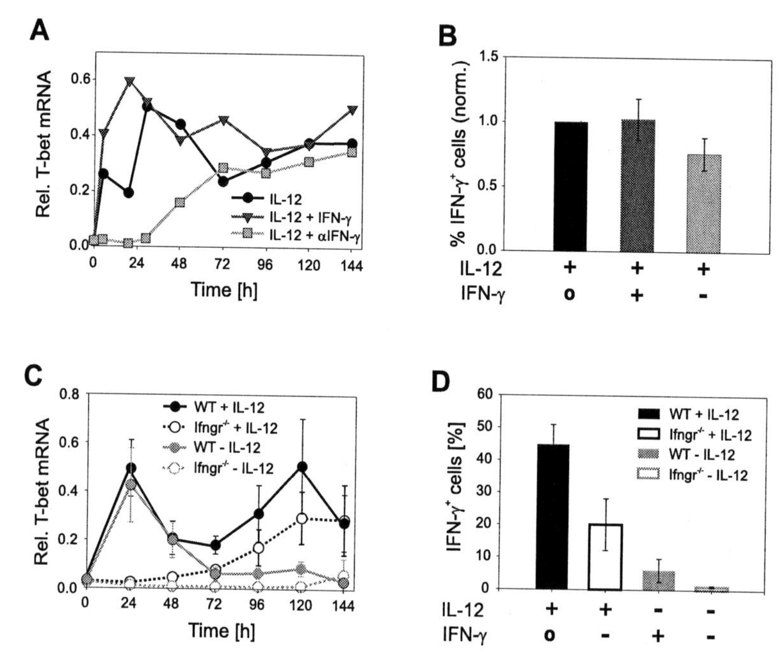

Figure 6.1. T-bet expression in the late phase controls cytokine memory for IFN-γ. **(A)** Naïve Th-cells ($CD4^+CD62L^{hi}CD44^{lo}$ purified by flow cytometry), isolated from Balb/c mice, were stimulated for six days with IL-12, and IFN-γ signaling remaining unperturbed (black, 'o'), with IL-12 and recombinant IFN-γ (dark grey, '+') or with IL-12 and blocking antibodies to IFN-γ (light grey, '-'). T-bet mRNA kinetics were quantified by RT-PCR. One representative example out of three independent experiments is shown. **(B)** Cells cultured as in (A) were re-stimulated with PMA and ionomycin after six days of culture and stained intracellularly for IFN-γ (raw data can be found in Fig. A.5 in Appendix A.4). Mean and s.d. of the frequency of IFN-γ producing cells under all three conditions was calculated from three independent experiments, normalized to conventional Th1-inducing conditions (IL-12). **(C)** Naïve Th-cells, isolated from C57BL/6 or *Ifngr*$^{-/-}$ mice, were stimulated with IL-12 (+IL 12) or with IFN-γ and blocking antibodies to IL-12 (-IL 12) and T-bet mRNA kinetics were quantified by RT-PCR. **(D)** After six days, cells were re-stimulated with PMA and ionomycin and stained intracellularly for IFN-γ. Mean and s.d. of the frequency of IFN-γ producing cells was calculated from four independent experiments.

blocking antibodies to interrupt IFN-γ signaling (Fig. 6.1A, light grey line). This perturbation resulted in a moderate decrease in the number of IFN-γ-producing cells in the recall response by ~20 % (Fig. 6.1B, light grey). Then, *Ifngr* deficient cells were analyzed, where T-bet expression was abolished in the first 48 hrs and expression was reduced in the late phase of priming (Fig. 6.1C, black dotted line). The polarization efficiency in these cells was reduced by ~50 % compared to wildtype cells (Fig. 6.1D). Finally, the role of the second wave of T-bet expression was investigated, which is IL-12 dependent. When cells were primed with IFN-γ in the absence of IL-12, T-bet expression in the first 48 hrs was left intact, while expression in the late phase was strongly reduced (Fig. 6.1C, solid grey line). Under these conditions, IFN-γ cytokine memory was reduced by ~80 % (Fig. 6.1D). No polarization occurred when *Ifngr* deficient cells were primed in the absence of IL-12 because T-bet was not expressed under these conditions (Fig. 6.1C+D, grey dotted line). These data suggest that the second, IL-12 dependent wave of T-bet expression controls IFN-γ expression in the recall response, while the first, IFN-γ dependent wave has a minor impact. Since in addition to controlling the first wave of T-bet expression, IFN-γ signaling enhances expression during the second wave, also *Ifngr* deficient cells exhibit reduced polarization efficiency (Fig. 6.1C+D). The presented results suggest that T-bet can drive Th1 differentiation only in the late phase of priming.

6.2 T-bet levels are strongly correlated with IFN-γ memory

To test the hypothesis that the expression of T-bet in the late phase controls the polarization efficiency, the correlation between T-bet expression and the frequency of IFN-γ-producing cells in the recall response was assessed. Data from five experiments (three of which are shown in Fig. 4.4) under Th1 and perturbation conditions was used. Cytokine memory for IFN-γ was uncorrelated with T-bet expression after 24 hrs (Fig. 6.2A), but highly correlated with its expression after 96 hrs, as shown by a correlation coefficient (cc) of 0.93 (Fig. 6.2B). By calculating the correlation coefficients for each time point during Th1 priming a strong correlation was found between 72 and 120 hrs (cc > 0.8), but only weak correlation between 0 and 48 hrs (c=0.2-0.5) (Fig. 6.2C). The correlation was maximal between 72 and 120 hrs, but decreased at 144 hrs, right before the recall response was induced. Since the weakly positive correlation coefficients between 0 and 48 hrs were associated with large confidence intervals, no significant correlation was found at these time points. By contrast confidence intervals were small in the late phase, giving support to the proposed strong positive correlation at that

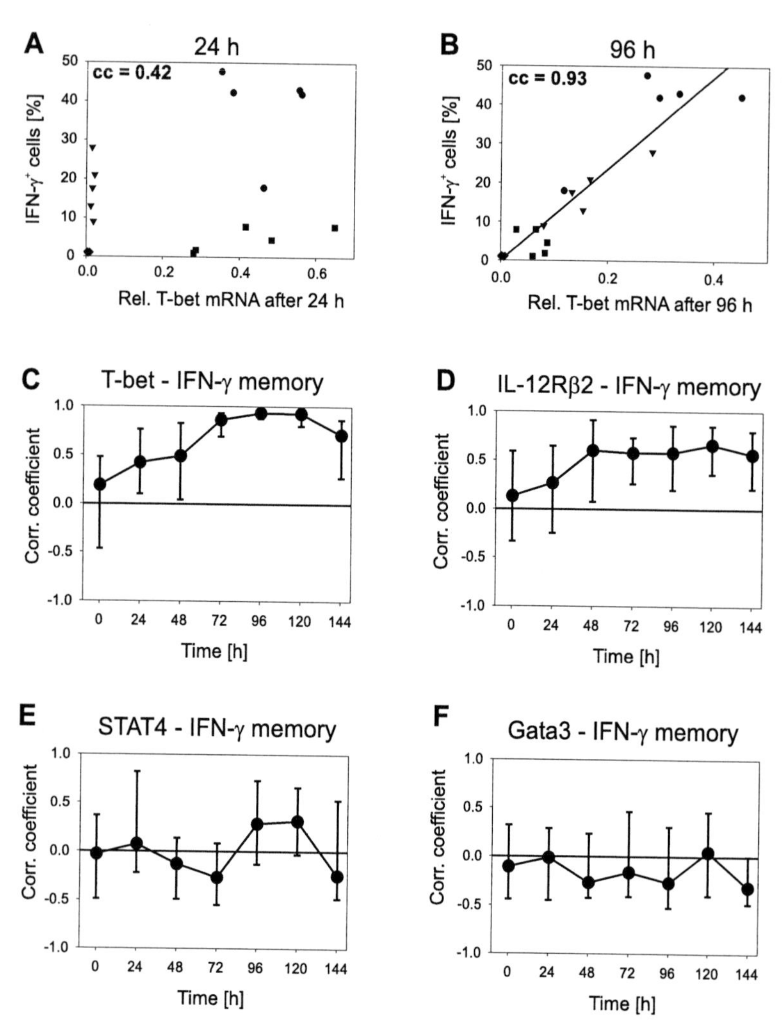

Figure 6.2. T-bet levels correlate with IFN-γ re-expression. (A,B) Naïve Th-cells isolated from C57BL/6 (circles, squares) or *Ifngr*$^{-/-}$ mice (triangles, diamonds) were stimulated with IL-12 (circles, triangles) or with IFN-γ and blocking antibodies to IL-12 (squares, diamonds). After six days of primary activation, cells were re-stimulated with PMA and onomycin and the frequency of IFN-γ producing cells was determined by intracellular cytokine staining. For five independent experiments, T-bet mRNA expression after 24 hrs (A) or 96 hrs (B) of stimulation is plotted against the frequency of IFN-γ producing cells in the recall response. The correlation coefficient (cc) is indicated and linear regression is shown for (B). **(C)** The correlation coefficient as indicated in (A) and (B) was determined for each time point. The error bars indicate a 95 % confidence interval, as estimated by bootstrap analysis. **(D-F)** A similar analysis was performed for IL-12Rβ2 expression (D), STAT4 (E) and GATA3 (F).

period. The presented results suggest that T-bet expression in the late phase is predictive for IFN-γ memory. Therefore, the second, IL-12-driven wave of T-bet expression seems to determine the polarization efficiency. Hence, the presented results uncover the molecular mechanism underlying the unique role of IL-12 in Th1 differentiation.

In the past years, there has been some debate about the question, whether T-bet or STAT4 is the main Th1 inducing transcription factor (Usui et al., 2003; Thieu et al., 2008). Since STAT4 activity is controlled by the expression level of IL-12Rβ2, also the correlation between the IL-12 receptor and IFN-γ re-expression was analyzed. Similar to T-bet, a significant positive correlation was observed between 72 and 144 hrs of primary activation (Fig. 6.2D). Such a positive correlation could result from a direct involvement of STAT4 in imprinting of the IFN-γ gene. Alternatively, it could be an indirect effect, arising because IL-12 signaling controlled T-bet expression in the late phase, which in turn is strongly correlated with IFN-γ memory (Fig. 6.2C). The latter possibility is supported by the fact that IL-12Rβ2 exhibits a much weaker correlation than T-bet, arguing for an indirect effect. Alternatively, the correlation might be weak, because STAT4 activity is not only controlled through the expression level of IL-12Rβ2. As STAT4 activity is not only regulated by phosphorylation, but also at the transcriptional level, the correlation between STAT4 mRNA and IFN-γ re-expression was analyzed, but no significant correlation was observed (Fig. 6.2E). Since also GATA3 has been implicated in the regulation of IL-12 signaling, it was tested, whether there was any relationship between GATA3 mRNA levels and IFN-γ memory. As shown in Fig. 6.2F, neither a positive, nor a negative correlation could be detected. Taken together, among the genes tested, only T-bet and IL-12Rβ2 expression levels (in the late phase) exhibit significant correlation with IFN-γ re-expression.

6.3 The concert of transcription factors active in the late phase

In the previous section, it was shown that T-bet expression in the late phase of primary activation is predictive for IFN-γ re-expression. But why is T-bet action restricted to this particular phase? It has been shown that T-bet does not act solitarily in Th1 differentiation, but that it cooperates with other transcription factors, such as Runx3, Hlx and STAT4 (Djuretic et al., 2007; Mullen et al., 2002; Thieu et al., 2008). Measuring the mRNA kinetics of Hlx and Runx3 during primary Th1 differentiation, both factors were found to be expressed at very low amounts during the first two days of priming and were upregulated between 72 and 120 hrs of stimulation (Fig. 6.3A+B).

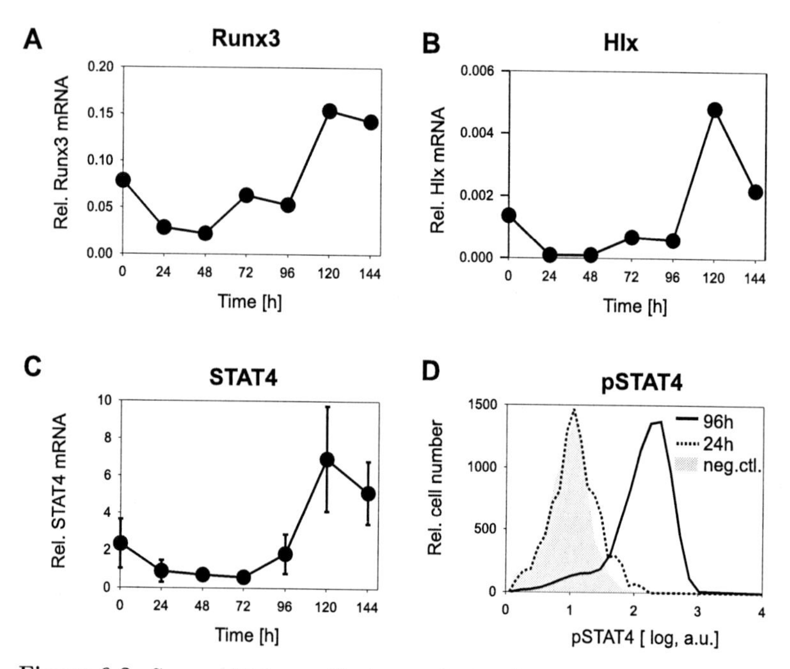

Figure 6.3. Several Th1-specific transcription factors are activated late during differentiation. Naïve Th-cells from C57BL/6 mice were cultivated under Th1-inducing conditions for six days. Runx3 (A), Hlx (B) and STAT4 (C) mRNA expression was quantified by RT-PCR. (D) At the indicated time points during stimulation, phosphorylated STAT4 was measured by intracellular staining and flow cytometry (solid lines). To assess background staining, cells stimulated in the absence of IL-12 were used (shaded area). One representative example out of three (Runx3) or two (Hlx, pSTAT4) independent experiments is shown. The error bars in (C) indicate s.d. of three independent experiments.

Similar kinetics were observed for STAT4 mRNA levels (Fig. 6.3C). Moreover, STAT4 phosphorylation was present in nearly all cells after 96 hrs of stimulation, while in the early phase (after 24 hrs) no signal was detected (Fig. 6.3D). Taken together, the presented data show that Runx3 mRNA, Hlx mRNA and phosphorylated STAT4, which is probably controlled by the STAT4 mRNA levels and by the kinetics of IL-12Rβ2 (see chapter 3.2) are present at high levels only in the late phase of priming. Therefore the availability of cooperating transcription factors might be one of the mechanisms restricting T-bet's ability to induce Th1 differentiation to the late phase of activation.

In this chapter, it has was shown that the expression level of T-bet in the late, not in the early phase of primary activation is predictive for IFN-γ memory. Since T-bet expression in the late phase requires IL-12 stimulation, this finding answers the long-standing question of why IL-12, but not IFN-γ can drive Th1 differentiation.

Chapter 7

Discussion

7.1 The core Th1 differentiation network

One major aim of this work was to identify the minimal network model that could account for the experimental observations during Th1 differentiation. This minimal model can be interpreted as the core of the network mediating the effect of IL-12 and IFN-γ in Th1 differentiation. In the presented work, this goal was approached in a systematic manner. A set of kinetic measurements of Th1 specific genes were generated under standard Th1 inducing conditions and with perturbation of all three key signals: IFN-γ, IL-12 and TCR-stimulation. Starting with a literature based model, new regulatory interactions were identified, so that the completed network model could account for the expression kinetics measured experimentally. Notably, previously unknown pathways were found, but it was not necessary to include additional genes. Statistical model comparison was used to show that the completed network was superior to the initial literature-based model in accounting for the measurements. In the completed two-loop model (Fig. 4.1), T-bet expression is controlled by two sequentially acting positive feedback loops, one mediated by IFN-γ the second one mediated by the IL-12/IL-12R pathway. Termination of the antigen signal acts as a switch between the T-bet/IFN-γ and the T-bet/IL-12R feedback loops, because TCR signaling promotes the former and inhibits the latter. These regulatory interactions between T-bet, IFN-γ, IL-12Rβ2 and the antigen-stimulus can explain the major features of the expression kinetics during primary activation, such as two-peaked T-bet expression with a single IFN-γ peak and delayed induction of IL-12Rβ2. Therefore the model described here seems to constitute the core of the gene network controlling Th1 differentiation.

The two-loop model describes the regulation of only three Th1 specific genes, but many more have been shown to be involved in the differentiation process. Since in the data set used all alternative differentiation pathways

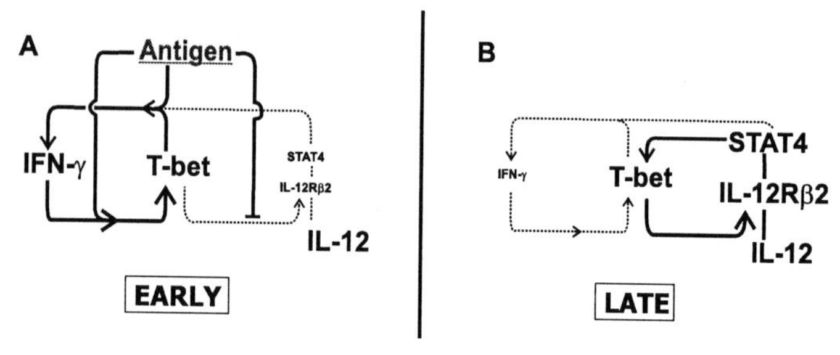

Figure 7.1. Inactive and active pathways early and late during differentiation. (**A**) In the early phase of priming (<48 hrs) T-bet and IFN-γ form a positive feedback loop, which requires antigenic stimulation. (**B**) In the late phase (>72 hrs) T-bet forms a positive feedback loop with IL-12/IL-12R, because antigen-dependent repression of the receptor is released. Solid lines denote high pathway activity, dotted lines low activity. The font size represents expression level of the gene.

were blocked, the number of active pathways was reduced. Nevertheless, possible network extensions were tested by including additional players, such as STAT4 and GATA3 that had been implicated in Th1 differentiation previously (Usui et al., 2003, 2006). These genes are indeed transcriptionally regulated during primary Th-cell activation, but the signals described by the two-loop model are not involved in their regulation. These results suggest that the gene-regulatory network active during Th1-cell activation and differentiation might have a modular structure (Hartwell et al., 1999). The module formed by T-bet and the Th1 inducing signals IL-12 and IFN-γ is described in this work. Other modules, involved in regulation of alternative fates or in proliferation, survival and cells death remain to be analyzed in a similar manner.

7.2 The early phase of differentiation

The analysis of the network revealed that the differentiation process can be divided in distinct phases. In the early phase, when TCR-signaling occurs, IL-12 signaling is low due to TCR-dependent repression of the IL-12Rβ2 subunit, and T-bet expression is induced by autocrine IFN-γ signaling. After termination of antigenic stimulation, the IL-12R is de-repressed and high-level IL-12 signaling occurs. Then, IL-12 acts to maintain T-bet expression at a high level. The regulatory interactions present in the early and the late phase of activation are depicted schematically in Fig. 7.1.

Until now, expression of T-bet had been considered to be regulated by IFN-γ and STAT1 (Afkarian et al., 2002; Lighvani et al., 2001). In chapter 3.1, data is presented, showing that IFN-γ indeed controls T-bet expression, but only in the first two days of activation, a period when also previous measurements were conducted. During this phase a positive feedback loop, where autocrine IFN-γ signaling induces T-bet, which in turn supports IFN-γ production, is active and controls T-bet expression (Fig. 7.1A). T-bet upregulation is observed rapidly within a few hours upon onset of stimulation (Fig. 6.1), because it reacts very sensitively even to low IFN-γ concentrations (Lighvani et al., 2001). While the presented data show clearly that the T-bet/IFN-γ feedback loop is the central regulatory mechanism in the early phase, it remains to be clarified, which mechanisms start up the feedback loop. One possibility, also assumed in the two-loop model, is that a low basal T-bet expression, already present in naïve cells, initiates the loop by upregulating IFN-γ expression as soon as TCR signaling is present. In an alternative mechanism that has been proposed by Locksley and colleagues, low amounts of IFN-γ, induced by TCR-signaling independent of IL-12/STAT4 and IFN-γ/STAT1, would be the initial event (Grogan et al., 2001). After approximately two days of primary activation, the T-bet/IFN-γ feedback loop is shut down, because it requires TCR-stimulation: T-bet can induce high-level IFN-γ expression only in synergy with TCR-dependent signals, and IFN-γ induces T-bet together with antigenic signaling (Lighvani et al., 2001). In the present study the question of which TCR dependent transcription factors induce T-bet expression has been addressed for the first time. As shown in chapter 3.1, calcineurin-dependent transcription factors, such as NFAT, are likely to mediate the TCR-dependent effect on T-bet expression. The regulation of NFAT will be analyzed in detail in the second part of this thesis.

While high-level IFN-γ signaling occurs specifically in the early phase, intracellular IL-12 signaling is low during this phase, even in the presence of high IL-12 concentrations in the cell's environment (Fig. 7.1A, dotted lines). In this work, the mechanism that limits STAT4 activity in the first days of activation was identified: Surprisingly, signals downstream of the TCR repress expression of the IL-12Rβ2 subunit, mediated by the calcineurin/NFAT pathway. Although a NFAT1-dependent silencer element had been identified previously in the human *IL12rb2* promoter (van Rietschoten et al., 2001), no NFAT1 binding could be detected to that site (which is also not conserved in the murine promoter) and also not to two other candidate binding sites that were identified based on sequence analysis. Therefore it remains open, whether NFAT1 exerts its repressive effect through binding to a not yet tested region of the *Il12rb2* gene, whether the effect is mediated by other calcineurin dependent factors, or whether NFAT regulates IL-12Rβ2 expression through an indirect mechanism. Another mechanism that might contribute to reduced STAT4 phosphorylation levels seen in the first days of activation is the down-regulation of STAT4 mRNA: Within 24 hrs of stimulation STAT4 levels drop to 30 % of the value seen in naïve cells. To what extent transcriptional regulation of STAT4 affects IL-12 signaling will be discussed in the next section.

In summary, the early phase is mostly controlled by IFN-γ signaling, because IL-12Rβ2 expression is inhibited. T-bet and IFN-γ form a positive feedback loop, which supports primary IFN-γ production. However, T-bet expression during that time window is insufficient to drive differentiation.

7.3 The late phase of differentiation

After termination of TCR stimulation (>48 hrs) repression of IL-12Rβ2 is released and its expression levels rise slowly over several days. In previous studies, IFN-γ as well as IL-12 itself have been identified as positive regulators of IL-12Rβ2 expression (Szabo et al., 1997; Smeltz et al., 2002). Moreover, T-bet can induce IL-12Rβ2 expression, even when expressed ectopically in the absence of IL-12 and IFN-γ signaling (Afkarian et al., 2002). Our finding that not only IFN-γ, but also IL-12 induces T-bet expression, provides a unifying mechanism for the previously observed effects of IL-12 and IFN-γ on IL-12Rβ2 expression. In the mathematical model developed in the presented work, the only positive regulator of IL-12Rβ2 is T-bet and this is sufficient to account for the expression kinetics under all conditions investigated (Fig. 7.1B).

As already mentioned above, STAT4 mRNA is down-regulated upon onset of stimulation. Similarly to IL-12Rβ2, this repression seems to be released in the later phase of priming, but only after 96-120 hrs of primary activation. The question arises whether the slow increase of STAT4 phosphorylation, observed in the later phase of priming, is controlled by de-repression of IL-12Rβ2 or STAT4 expression. Analysis of the data presented here, shows that STAT4 becomes phosphorylated as soon as IL-12Rβ2 expression starts to increase, at time points when STAT4 mRNA expression remained at a constant low level (48-72 hrs). Therefore availability of IL-12Rβ2, not STAT4 seems to limit IL-12 signaling in the early phase of priming.

One key finding presented in this work is the fact that, in the late phase of primary Th-cell activation, IL-12 directly controls T-bet expression and STAT4 is bound to the *Tbx21* enhancer. Because IL-12 maintains T-bet expression, which in turn induces expression of IL-12Rβ2, IL-12 enhances indirectly expression of its own receptor and T-bet indirectly promotes its own expression by inducing IL-12Rβ2 (Fig. 7.1B). Through this positive feedback loop, IL-12 maintains sustained T-bet expression in the late phase of primary Th-cell activation, which is required for establishing IFN-γ cytokine memory. This mechanism links IL-12, the major Th1-inducing cytokine, and T-bet, the Th1 master transcription factor.

Although effects of IFN-γ are more obvious in the early phase of priming, a small IFN-γ induced increase of T-bet expression levels was also observed in the late phase. In the two-loop model this effect was attributed to IFN-γ signaling in the early phase, which could accelerate start-up of the IL-12R/T-bet feedback loop in the late phase. If T-bet and IL-12Rβ2 are initially expressed at low amounts, mutual induction is inefficient and thereby proceeds with slow kinetics. This slow induction of IL-12Rβ2 and T-bet is accelerated significantly, when T-bet expression is induced by IFN-γ in the early polarization phase. An alternative explanation for the IFN-γ effect in the late phase is discussed on chapter 5.2. Several observations suggest that the IFN-γ/T-bet feedback loop remains active also in the late phase of priming, albeit at low levels (Fig. 7.1B, dotted lines). Surprisingly, low levels of IFN-γ are expressed in the entire population also in the late phase of culture, but only in the presence of IL-12. This observation suggests that, when the IL-12R/T-bet feedback loop is active, T-bet expression is further increased through continuous autocrine IFN-γ signaling. Although the presented data support this hypothesis, it remains to be directly tested experimentally. In general, our observations suggest that Th1-cells possess two different modes of IFN-γ expression: (1) Low-level expression that is induced in a uniform manner in the whole population and that is sufficient for autocrine signaling, while (2) high-level expression that is induced transiently by TCR-stimulation, is typically only seen in a fraction of cells and mediates paracrine effector functions of IFN-γ.

7.4 The Th1 regulatory network in its physiological context

In vivo, naïve Th-cells are activated in the context of a physical interaction with an antigen-presenting cell (APC) in the lymph node, which stimulates the TCR and produces IL-12 (schematic representation in Fig. 7.2). Initially, IL-12 production in the APC is induced by bacterial products and TCR-stimulation elicits IFN-γ production by the T-cell. Then IL-12 expression is further enhanced by T-cell derived signals such as IFN-γ and CD40-stimulation (Koch et al., 1996; Trinchieri, 2003). As shown in red in Fig 7.2, the synergistic activation of IFN-γ production by APC-derived IL-12 together with antigen and (IFN-γ induced) T-bet will close a positive

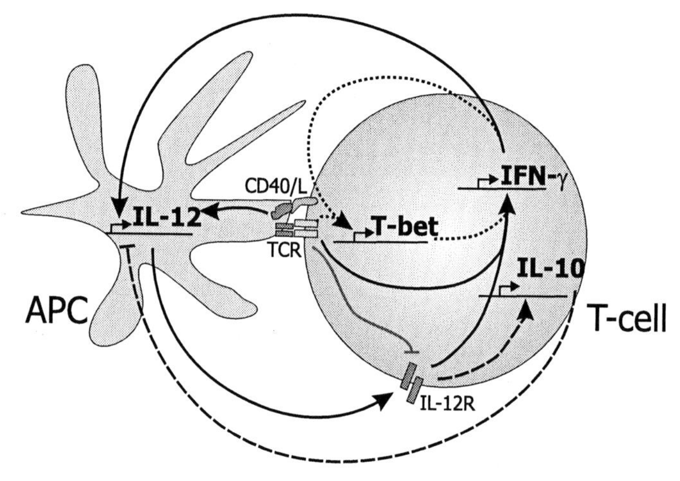

Figure 7.2. Autocrine and paracrine feedback loops during the APC-T-cell interaction. In an autocrine positive feedback loop (dotted arrows) IFN-γ induces T-bet, which in turn promotes IFN-γ transcription. A paracrine positive feedback loop (solid arrows) between APC and T-cell is formed, when T-cell derived IFN-γ induces IL-12 expression in the APC, in synergy with CD40-stimulation, and IL-12 promotes IFN-γ transcription in the T-cell, in synergy with TCR stimulation. Negative paracrine feedback regulation (dashed arrows) is mediated by IL-10, which is induced by IL-12 in the T-cell and inhibits IL-12 expression in the APC. Negative feedback regulation is prevented by TCR-dependent repression of IL-12R expression in the T-cell (grey line).

feedback loop between APC and T-cell. Therefore the early wave of T-bet expression is important for IL-12 induction in the APC.

During interaction with the APC, the T-cell responds only weakly to IL-12, because TCR-dependent signals suppress IL-12Rβ2 expression. But what could be the physiological function of suppressing signaling downstream of a central Th1-inducing signal during T-cell activation? One reason could be to maintain plasticity in the early phase of activation to prevent an erroneous lineage decision. In addition, repression of IL-12Rβ2 expression might be important to prevent IL-10 dependent negative feedback regulation of IL-12 (Fig. 7.2, dashed arrows). IL-12 is involved in termination of the immune response by inducing expression of the anti-inflammatory cytokine IL-10 in Th1-cells (Chang et al., 2007; O'Garra and Vieira, 2007). IL-12 production itself is inhibited by IL-10, which thereby acts as a negative feedback regulator (O'Garra and Vieira, 2007). T-cells might be particularly competent for IL-10 production during the APC-T-cell interaction, because IL-10 expression seems to require concomitant stimulation of Notch by Notch-ligands expressed on the APC-surface (Rutz et al., 2008). Inhibition of IL-12 signaling during TCR stimulation could therefore be important to delay IL-10 production by Th1-cells (Assenmacher et al., 1998), thereby preventing instant negative feedback regulation of T-cell activation.

From our finding that TCR-stimulation inhibits IL-12R expression another question arises: What is the source of IL-12 in the late phase, when interaction with the APC has been terminated, but when IL-12 is crucial to maintain T-bet expression in order to induce differentiation? As long as the T-cell still resides in the lymph node, it can probably respond to the IL-12 that the APC's have produced during interaction with the T-cells. Soon after their priming, however, Th1-cells leave the lymph node and migrate to the site of inflammation. Here they exert their effector functions, such as activation of macrophages by IFN-γ. IFN-γ also induces IL-12 production in macrophages (Trinchieri, 2003), which could be required to maintain the second wave of T-bet expression. In this context, also the low-level IFN-γ expression that was observed in the late phase of priming, might be important to induce IL-12 expression in macrophages also in the absence of antigenic stimulation. Some studies *in vivo* have already suggested that continuous IL-12 signaling is required for an effective Th1 response (Yap et al., 2000; Stobie et al., 2000).

7.5 Stabilization of the Th1 phenotype

In chapter 6 it was shown that the frequency of IFN-γ producing cells in a recall stimulation is regulated by the expression of T-bet in the late phase of primary activation (>48 hrs) in a dose-dependent manner. Importantly, this T-bet dose-dependence of Th1 priming held true for different stimulation protocols (e.g. with or without IFN-γ signaling), strongly suggesting that T-bet expression in a specific time window, rather than other factors, is the decisive event. Which mechanisms might restrict T-bet dependent Th1 differentiation to the late phase of priming? One temporal constraint on Th1 priming has been shown previously, in that memory expression of the *Ifng* gene requires entry into the S-phase of the first cell cycle (Bird et al., 1998; Richter et al., 1999). However, as this occurs at ~20 hrs of stimulation while the T-bet effect is strongest at about 96 hrs, S-phase entry seems to be only one of a number of necessary events that allow or mediate T-bet action. In agreement with previous reports, it was shown that Runx3 and Hlx, which are induced by, and cooperate with T-bet were specifically expressed in the late phase of primary activation (Mullen et al., 2002; Djuretic et al., 2007). Moreover, STAT4 activation is also restricted to the late phase and also cooperates with T-bet on the induction of multiple Th1 specific genes (Thieu et al., 2008). In addition, it has been reported recently that STAT5-dependent induction of chromatin accessibility, occurring between 24 and 48 hrs of primary activation, is required for binding of T-bet to the *Ifng* locus (Shi et al., 2008). These results suggest that T-bet acts in concert with other transcription factors, such that co-factors restrict T-bet's ability to imprint the *Ifng* gene to the late phase of Th1 differentiation.

The observed correlation between T-bet levels in the late phase of priming and IFN-γ expression in the recall stimulation could reflect a direct causal effect of T-bet on Th1 differentiation. Alternatively such a correlation would also be observed, if T-bet expression and Th1 differentiation were controlled by the same upstream factor. Such a common regulator could be STAT4 that controls T-bet transcription in the late phase (section 3.1) and induces epigenetic changes on the *Ifng* locus (Chang and Aune, 2005; Zhang and Boothby, 2006). If this was the case, the correlation between the differentiation efficiency and IL-12Rβ2 expression that controls STAT4 activity, would be expected to be at least as strong as the correlation between the differentiation efficiency and T-bet levels. However, the presented analyses show that, although the differentiation efficiency is correlated with IL-12Rβ2 expression (cc~0.6), the correlation coefficient is much lower than for T-bet (cc~0.9). Therefore, T-bet seems to directly control Th1 differentiation and IFN-γ re-expression.

It is still unclear, whether T-bet needs to be expressed only transiently to induce differentiation or whether it must be present constitutively to stabilize the Th1 phenotype. T-bet is clearly required to initiate Th1 differentiation in that histones are modified and chromatin is remodeled in regulatory regions on the *Ifng* gene (Chang and Aune, 2005; Hatton et al., 2006; Shnyreva et al.,

2004), but it seems to be partially dispensable at later stages of differentiation (Mullen et al., 2002). In mature Th1-cells the phenotype might be stabilized on the epigenetic level. Nevertheless, T-bet is also expressed in resting and memory Th1-cells and it is still unclear how expression is maintained, because the present analysis reveals a strict dependence of T-bet expression on extracellular signals. It could be shown that in the absence of antigenic signals IL-12 is required to maintain high T-bet expression. However, at least *in vitro* differentiated Th1-cells that regularly receive antigenic stimulation remain competent for IFN-γ expression also in the absence of IL-12. By contrast, some studies *in vivo* suggest that continuous IL-12 signaling is required to maintain Th1 memory (Yap et al., 2000; Stobie et al., 2000). Therefore, long-term *in vitro* Th1 cultures, where the cells receive a weekly antigenic stimulation, might not reflect accurately the situation *in vivo* and might underestimate the role of IL-12. Alternatively, the low T-bet levels that can be expressed in the absence of IL-12 might be sufficient for Th1 phenotype stabilization. However, if indeed autocrine IFN-γ signaling was responsible for the T-bet plateau in the late phase of priming, as suggested by some observations, either IL-12 would be required to induce IFN-γ or IFN-γ must be produced by other cells. In any case, T-bet expression would depend on extracellular signals. Therefore, future studies are required to reveal, whether IFN-γ or alternative, possibly cell intrinsic mechanisms, such as T-bet auto-activation, are responsible for the observed low T-bet plateau in the late phase of priming in the absence of IL-12.

7.6 Summary and outlook

A new model of Th1 priming is presented in this study supported by kinetic experiments and mathematical modeling. Th1 differentiation is induced in a two-step process controlled by two sequentially acting positive feedback loops. In the early phase, when the TCR is stimulated, the Th1 master transcription factor T-bet is induced by a positive feedback loop mediated by IFN-γ. During that phase, expression of the IL-12 receptor β2 subunit is repressed by TCR signaling. Therefore, IL-12Rβ2 is induced by T-bet only after termination of the antigen stimulus. This allows activation of STAT4 by IL-12, which is required to maintain T-bet expression in the late phase of activation, when it is drives Th1 differentiation.

Since this work is the first attempt to develop a data-based model of the Th1 differentiation network, a number of questions remain to be addressed in the future. Previously unknown regulatory interactions were described, but some of the underlying molecular mechanisms remain unclear. The antigen-dependent repression of IL-12Rβ2 is most likely mediated by NFAT transcription factors, but no direct binding of NFAT to the IL-12Rβ2 promoter could be detected. Therefore, NFAT either represses IL-12Rβ2 through an indirect mechanism or it acts on regulatory sites that were not tested here. Notably, it is also unclear how T-bet induces IL-12Rβ2, this as well could be an indirect effect. Also some aspects of T-bet regulation remain puzzling. IFN-γ induced STAT1 and IL-12 dependent STAT4 bind to the same enhancer on the *Tbx21* locus (Yang et al., 2007). But why can STAT1 only act together with NFAT, while STAT4 can act independently of TCR-signals?

As already discussed in the previous section, the mechanisms required for stabilization of the Th1 phenotype remain ill defined. Future experiments should reveal, whether T-bet can be expressed in the absence of IL-12 and antigenic signals. To address this question, IL-12 and IFN-γ should be removed in the late phase of priming, or, alternatively, T-bet expression in the second week of culture without IL-12 stimulation could be analyzed.

Finally, the theoretical analysis of the transcriptional network in Th-cells will hopefully be extended to other modules in the future. Of particular interest would be the interaction between the regulatory network that controls differentiation toward the different lineages. In this context, the surprising finding that the expression levels of STAT4, as part of the Th1 differentiation program, and GATA3, a Th2-specific transcription factor, are highly correlated, should be analyzed further. To better understand the situation *in vivo*, we should work towards a better understanding of how cell make differentiation decisions based on a complex mixture of signals.

It became clear from the presented analysis that the transcription factor NFAT1, activated by TCR stimulation, plays an important role in Th1-differentiation and activation. It is not only required to induce high-level expression of the Th1 effector cytokine IFN-γ, but it also drives expression of the Th1 master transcription factor T-bet and, most surprisingly, inhibits expression of the IL-12 receptor β2 subunit. In the second part of this thesis the regulation of NFAT1 activity will be analyzed in some detail.

Part II

The NFAT signaling network

Chapter 8

Introduction

In the first part of this thesis, antigen signaling was introduced as one central stimulus driving Th1 differentiation. Through activation of NFAT transcription factors antigenic signaling controls expression of T-bet and IL-12Rβ2. In the second part of this work, the regulation of NFAT activity will be analyzed in detail, again with a combination of experimental measurements and mathematical modeling.

Nuclear factor of activated T-cells (NFAT) was first identified as an important regulator in the immune system, but it also plays a role in other processes such as differentiation of skeletal muscle cells and osteoclasts (Hogan et al., 2003). The NFAT family consists of five members (NFAT1-NFAT5), four of which are activated by calcium signaling (NFAT1-NFAT4). These four NFAT factors fulfill over-lapping yet distinct functions and share most of their regulators. In this project, NFAT1, also known as NFATp or NFATc2, was investigated and is hereafter called NFAT.

8.1 Regulation of NFAT1

All NFAT proteins are activated by calcium signaling. Stimulation of the TCR results in an increase of the cytoplasmic calcium concentration due to the opening of store-operated calcium channels (Liu, 2009). Up to four Ca^{2+} ions bind to the cytoplasmic calcium sensor calmodulin and induce a conformational change that allows calmodulin to bind the phosphatase calcineurin. This interaction activates the phosphatase activity of calcineurin by inducing dissociation of an auto-inhibitory domain from the active site (Liu, 2009). Calcineurin then activates NFAT by dephosphorylating multiple serine residues in its regulatory domain. As shown schematically in Fig. 8.1, NFAT consists of four functionally distinct domains: the N-terminal transactivation domain, the regulatory domain, the DNA-binding domain and a C-terminal domain involved in dimer formation (Hogan et al., 2003).

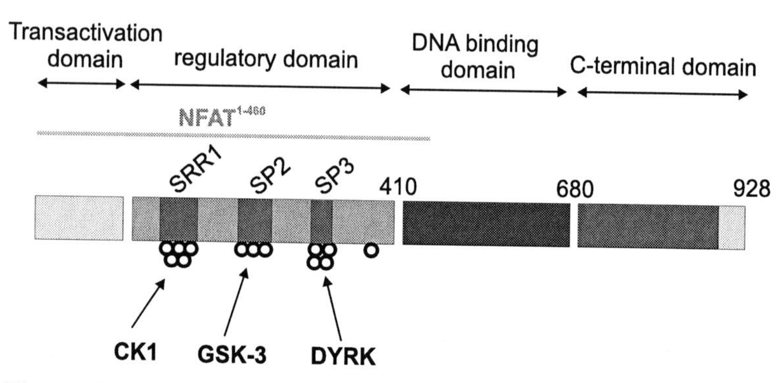

Figure 8.1. The domain structure of NFAT1. The domain structure of murine NFAT1 is shown. In the regulatory domain, three different phosphorylation motifs are indicated (SRR1, SP2, SP3), each consisting of several phosphorylated serine residues (circles). The kinases (CK1, GSK-3, DYRK) responsible for phosphorylating these motifs are also shown. In gray, the $NFAT^{1\text{-}460}$ construct used in the present study is shown. Scheme was adapted from Hogan et al. (2003).

NFAT contains 13 conserved phosphorylation sites (Fig. 8.1, circles) in its ~300 amino acid long regulatory domain that are phosphorylated in resting cells and become dephosphorylated by calcineurin in response to TCR stimulation. The phosphorylation status of these 13 serine residues controls NFAT's transcriptional activity through regulating its subcellular localization and its affinity for DNA. It has been proposed that fully phosphorylated NFAT resides predominantly in the cytoplasm, because a not yet identified nuclear export signal (NES) is exposed, while the nuclear import signal (NLS) is masked (Okamura et al., 2000). Dephosphorylation by calcineurin is thought to induce a conformational change that unmasks the NLS and covers the NES, resulting in nuclear accumulation. NFAT's phosphorylation state does not only regulate its subcellular localization, but also controls its DNA binding properties (Neal and Clipstone, 2001). In resting cells, where NFAT is fully phosphorylated, DNA binding activity on NFAT target sites is low, but increases when NFAT is dephosphorylated upon stimulation (Shaw et al., 1995; Okamura et al., 2000). In summary, dephosphorylation controls NFAT function by inducing import into the nucleus and by increasing the binding affinity for DNA.

In resting cells, the phosphorylated state of NFAT is maintained by constitutively active kinases that are also responsible for NFAT deactivation, when the stimulus is removed. Interestingly, dephosphorylation of all 13 residues is catalyzed by a single phosphatase calcineurin, but multiple kinases are involved in rephosphorylation, each targeting specific residues. Casein kinase 1

(CK1) targets a serine rich regions (SRR1), which contains five phosphorylatible residues (Fig. 8.1) (Zhu et al., 1998; Okamura et al., 2004). This kinase resides in a stable, high-affinity macro-molecular complex with NFAT in resting cells and has been proposed to act as a maintenance kinase, required to keep NFAT in the phosphorylated, cytoplasmic state in the absence of stimulation (Okamura et al., 2004). Glycogen synthase kinase 3 (GSK-3) controls NFAT activity by targeting a serine-proline repeat (SPxx) motif, named SP2, which contains three phosphorylated serine residues (Okamura et al., 2004; Beals et al., 1997b). Since dephosphorylation of SP-motifs is required for binding to DNA, GSK-3 directly controls transcriptional activity (Neal and Clipstone, 2001). Interestingly, GSK-3 is also regulated during antigenic stimulation: It is phosphorylated by the Akt kinase which is activated by CD28-co-stimulation and thereby inhibits GSK-3's enzymatic activity (Cross et al., 1995; Diehn et al., 2002). Therefore, co-stimulation increases NFAT activation by inhibiting a NFAT kinase. Notably, GSK-3 cannot act on completely dephosphorylated NFAT protein (Beals et al., 1997b). Instead, it requires a priming kinase, such as "protein kinase A" or "dual-specificity tyrosine-phosphorylation regulated kinase" (DYRK) (Beals et al., 1997b; Gwack et al., 2006; Arron et al., 2006; Sheridan et al., 2002). DYRK has been shown to phosphorylate another SPxx-repeat motif (SP3) (Neal and Clipstone, 2001; Gwack et al., 2006; Arron et al., 2006) and has been suggested to control export and transcriptional activity of NFAT. Furthermore, also mitogen-activated kinases, such as p38 have been implicated in NFAT regulation (Gómez del Arco et al., 2000). In summary, a number of kinases affect NFAT activity, potentially allowing it to integrate a multitude of signals. Differential regulation by kinases might also allow cell-type specific regulation of NFAT activity.

Several lines of evidence suggest that NFAT cycles continuously between nucleus and cytoplasm, in resting cells and also upon stimulation (Fig. 8.2). In resting cells, inhibition of nuclear export through leptomycin B results in accumulation of NFAT in the nucleus (Okamura et al., 2000). Since this effect is also observed, when phosphatase activity is blocked completely, fully phosphorylated NFAT seems to be able to enter the nucleus, albeit with a slow rate. Upon stimulation, NFAT is dephosphorylated by calcineurin and can then enter the nucleus within a few minutes (Beals et al., 1997a; Zhu et al., 1998). When calcineurin activity is inhibited, NFAT is rapidly rephosphorylated and translocates back to the cytoplasm (Timmerman et al., 1996; Okamura et al., 2000). These observations suggest that NFAT cycles continuously between the phosphorylated and the dephosphorylated state and between the nucleus and the cytoplasm (Fig. 8.2). In contrast to other transcription factors such as NFκB that are slowly deactivated (associated with *de novo* synthesis of regulatory proteins), NFAT is rapidly deactivated so

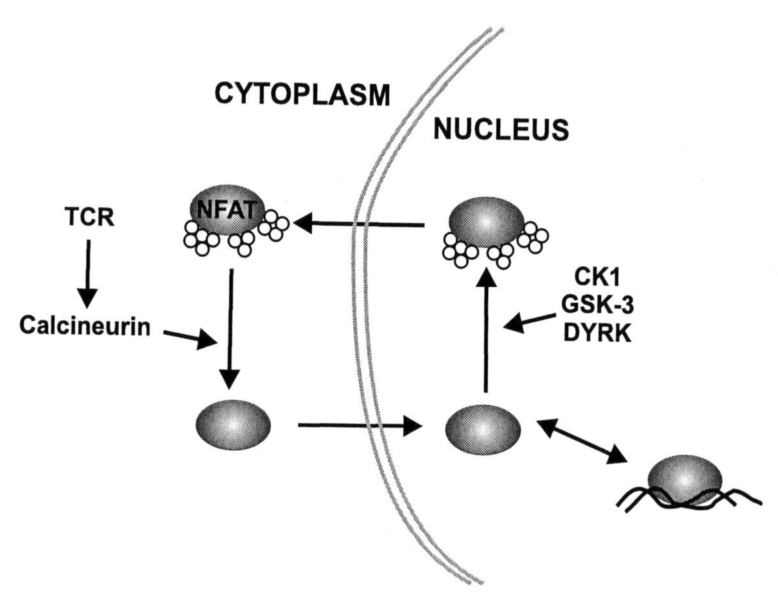

Figure 8.2. NFAT regulation. Upon TCR stimulation, calcineurin is activated and dephosphorylates NFAT. Dephosphorylated NFAT can then enter the nucleus, where it binds to DNA and induces target gene transcription. Different kinases, such as CK1, GSK-3 and DYRK participate in NFAT rephosphorylation. Phosphorylated NFAT can then be transported back to the cytoplasm.

that it can sense changes in the activation state of membrane-bound receptors. In such a system, the amount of transcriptionally active NFAT in the cell is controlled by the balance between kinase and phosphatase activity and by the resulting balance of import and export.

8.2 Objectives and experimental design

The fact that a number of different kinases are involved NFAT regulation suggests that they might play an important functional role. For example, different members of the NFAT family, all of which are controlled by calcineurin, show under certain conditions differential localization within the same cell (Abbott et al., 1998; Loh et al., 1996). Since not all phosphorylation motifs are completely conserved across the NFAT family, differential regulation by deactivating kinases might account for this observation. Moreover, the regulation of kinase activity might allow integration of multiple signals on the level of signal transduction. Therefore, the roles played by the kinases and the phosphatase that control NFAT activation were addressed in this project through a combination of experimental measurements and mathematical modeling.

Because NFAT cycles continuously between nucleus and cytoplasm and between the different phosphorylation states, the steady-state levels of transcriptionally active NFAT in the nucleus depend on the balance between these processes. Therefore, the transport and (de)phosphorylation rates were quantified, to gain a better understanding of how NFAT activity is regulated. In particular the following questions were addressed:

- Which is the mode of de- and rephosphorylation? Is there a limiting step or cooperativity?
- What are the transport properties of the partially phosphorylated states? Is a specific motif/kinase responsible for import or export?

To answer these questions, a mathematical model of the regulatory network controlling NFAT activity was developed and the rate constants of the phosphorylation and transport reactions were quantified from experimental data. Kinetic measurements of NFAT activation and deactivation were performed and localization and phosphorylation were monitored simultaneously. In addition, dose-response curves were measured, where calcineurin activity was titrated by using different amounts of the calcium ionophore ionomycin. To perturb the kinase activities, the pharmacological inhibitors LiCl (for GSK-3) and CKI7 (for CK1) were used. Since no pharmacological inhibitors are available for DYRK kinases, they were not explicitly included in the study.

Analysis of the model parameters and control theory was then used to understand the functional roles of the different enzymes. In contrast to the model described in part I, here the structure of the network was already known, and the project aimed at a quantitative understanding of the central processes.

Chapter 9

Experimental measurements

The experiments described in this chapter were performed together with Luca Mariani in the laboratory of Anjana Rao at Harvard Medical School.

9.1 Experimental system

To analyze of role of kinases in NFAT regulation, an experimental system was used that allowed simultaneous measurement of NFAT phosphorylation and localization. Because the cytoplasm of T-cells is small compared to the nucleus, it is difficult to quantify, which fraction of NFAT resides in each compartment. Therefore, a Hela cell line was used for all experiments, where the cytoplasm is of a similar size as the nucleus and can be easily identified (Appendix B.2). A cell line was used that was stably transfected with a murine NFAT1$^{1\text{-}460}$ construct, fused to "green fluorescent protein" (GFP) (Aramburu et al., 1998). This construct contained the regulatory domain, but missed the DNA binding region (Fig.. 8.1, gray).

To assess subcellular localization of NFAT, GFP fluorescence was measured with microscopy, followed by automated image analysis (see Appendix B.2). To define the area of nucleus and cytoplasm, the cells were stained with DAPI and fluorescently-labeled phalloidin. DAPI detects DNA in the nucleus and phalloidin is a natural cyclic peptide that interacts with high affinity with F-actin, which is restricted to the cytoplasm. To estimate the mean nuclear fraction of NFAT-GFP in the population, several hundred cells were analyzed for each data point through automated image analysis with the software CellProfiler (Carpenter et al., 2006). Nucleus and cytoplasm of each cell were identified through DAPI and phalloidin staining, respectively, and the integrated GFP intensity in both compartments was used to calculate the nuclear fraction. To assess NFAT's phosphorylation status, a change in the protein's electrophoretic mobility was exploited that is induced by dephosphorylation of the SP2 and SP3 motifs (Okamura et al.,

2000) and that results in a shift to a lower apparent molecular weight in the immunoblot (Fig. 9.1A). Immunoblot quantification allowed determination of the dephosphorylated fraction with high accuracy (see Appendix B.1).

9.2 Kinetics of NFAT activation and deactivation

In vivo, NFAT is activated by Ca^{2+} influx upon TCR stimulation. In cell culture, a rise in the cytosolic Ca^{2+} concentration can be elicited by the Ca^{2+} ionophore ionomycin. To terminate NFAT activation, the calcineurin inhibitor cyclosporine A (CsA) was used, which blocks activation of calcineurin by calmodulin. To measure activation and deactivation kinetics of NFAT, $\text{Hela}^{\text{NFAT-GFP}}$ cells were stimulated with 2 µM ionomycin for 20 min, then 2 µM CsA was added (experimental setup in Fig. 9.1A). Within minutes after addition of ionomycin, NFAT became dephosphorylated (Fig. 9.1B+C) and entered the nucleus (Fig. 9.1D). Compared to the dephosphorylation kinetics ($t_{1/2}$~2 min), rephosphorylation in response to CsA addition proceeded with much slower kinetics ($t_{1/2}$~40 min, Fig. 9.1C). Moreover, the delay between rephosphorylation and nuclear export ($t_{1/2}$~10 min) was considerably longer than the delay between dephosphorylation and nuclear import during the activation phase ($t_{1/2}$~2.5 min, Fig. 9.1, compare C and D). The maximal fraction of dephosphorylated NFAT varied strongly between experiments, but remained always well below 100 %. Similarly, the maximal nuclear fraction never exceeded 60-70 %. Surprisingly, even in resting cells, about 30 % of GFP fluorescence was detected in the nucleus, suggesting that also completely phosphorylated NFAT could be imported into the nucleus with reasonable efficiency.

9.3 Response to kinase inhibitors

To analyze the roles played by the kinases GSK-3 and CK1, their activity was perturbed by specific pharmacological inhibitors: LiCl blocks GSK-3, while CKI7 acts on CK1. To analyze the effect of kinase inhibition on the deactivation kinetics, cells were pre-incubated with LiCl (black triangles), CKI7 (grey circles) or both together (grey triangles) and stimulated with 2 µM ionomycin for 20 minutes (Fig. 9.2A,B). Phosphorylation and transport kinetics upon CsA addition were measured. A strong effect was observed in cells pre-treated with CKI7 (grey): Phosphorylation levels were increased ~3-fold in stimulated cells (Fig. 9.2A), while the nuclear fraction was increased ~2-fold (Fig. 9.2B). Moreover, CKI7 also resulted in a small, but significant

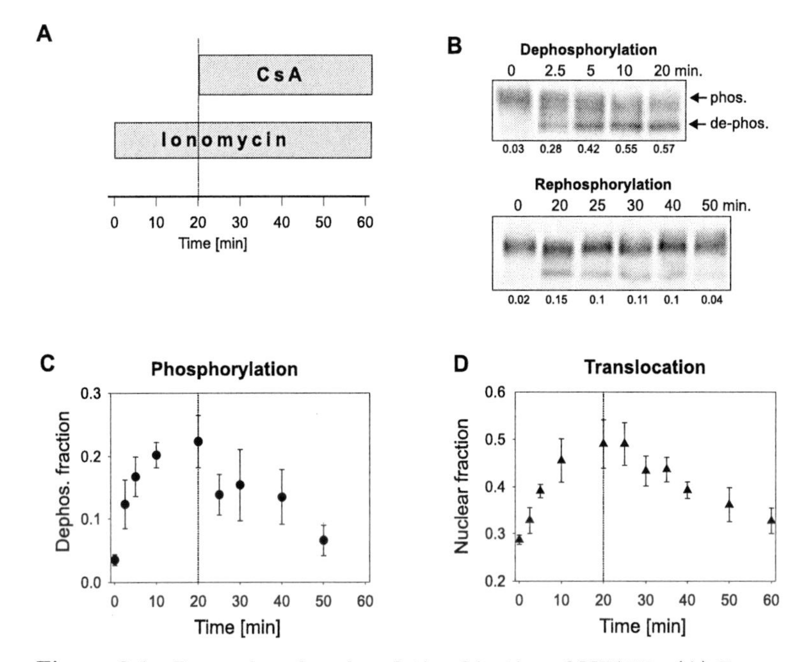

Figure 9.1. De- and rephosphorylation kinetics of NFAT1. (A) Experimental setup of kinetic measurements: Hela$^{\text{NFAT-GFP}}$ cells were stimulated with 2 µM ionomycin for 20 min, then 2 µM CsA was added. **(B)** At different time points during the activation phase (Dephosphorylation) and during the deactivation phase (Rephosphorylation), NFAT's phosphorylation status was analyzed by immunoblot (quantification of the dephosphorylated fraction is stated below each lane). **(C)** The experiments shown in (B) were performed three times and normalized to the average values detected in stimulated (20 min) and resting cells. **(D)** During the experiments described in (C), the nuclear fraction was measured at the indicated time points. Mean and s.d. are plotted.

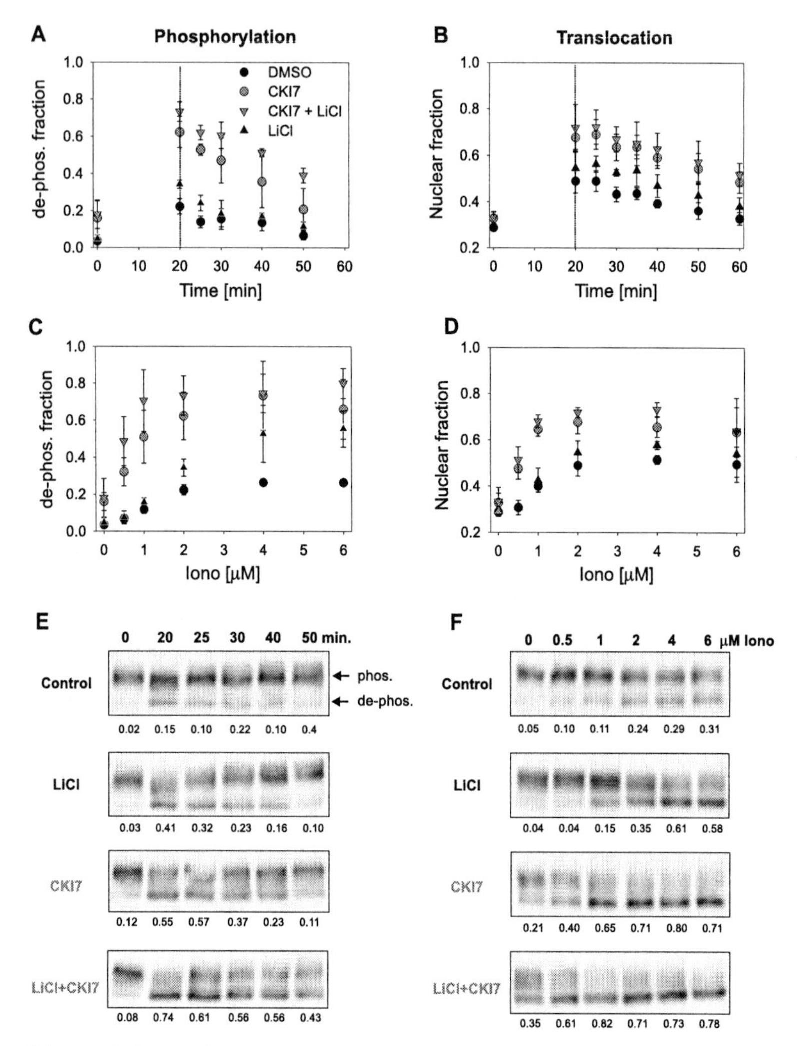

Figure 9.2. NFAT activation is increased by kinase inhibitors. (A,B) Hela$^{\text{NFAT-GFP}}$ cells were pre-incubated for 20 min with the GSK-3 inhibitor LiCl (black triangles), with the CK1 inhibitor CKI7 (grey circles) or with both together (grey triangles), and were then stimulated with 2 μM ionomycin. After 20 min CsA was added to the medium and NFAT rephosphorylation (A) and translocation (B) kinetics were analyzed. **(C,D)** Hela$^{\text{NFAT-GFP}}$ cells were pre-incubated with kinase inhibitors as in (A, B) and stimulated for 20 min with different ionomycin concentrations. Phosphorylation (C) and translocation (D) was quantified. Mean and s.d. of three independent experiments is shown, normalized to the mean values in resting (0 min, 0 μM) and stimulated (20 min, 2 μM) cells in presence of the respective inhibitors. (E, F) A representative western blot experiment for (A) and (C).

increase in the dephosphorylated and nuclear fractions in resting cells prior to stimulation (Fig. 9.2A+B, 0 min). LiCl had no detectable effect on resting cells, but induced a small increase in the dephosphorylated fraction and in the nuclear NFAT levels upon stimulation (Fig. 9.2A,B, compare black triangles and circles). Simultaneous treatment with LiCl and CKI7 showed that the inhibitor effects were additive (Fig. 9.2A+B, grey triangles).

In the next step, it was analyzed how the inhibitors affected phosphorylation and translocation, when the stimulus strength was varied. Cells, pre-treated with the inhibitors, were incubated with different ionomycin concentrations for 20 min and phosphorylation and translocation was quantified (Fig. 9.2C+D). In the absence of inhibitors the dose-response curve had a slightly sigmoidal shape that saturated at ionomycin concentrations above 2 µM (Fig. 9.2C+D, black circles). Interestingly, the effect of the kinase inhibitors differed between low (≤1 µM) and high ionomycin concentrations (≥2 µM). At low stimulus strength, addition of LiCl had no detectable effect, while CKI7 resulted in a 3-5-fold increase in the dephosphorylated and nuclear fractions (Fig. 9.2C+D). When LiCl was added together with CKI7, this strong CKI7-dependent increase was further enhanced (Fig. 9.2C+D). At an intermediate stimulus strength (2 µM) that was also used for the kinetic experiments above, the inhibitor effects were additive. At ionomycin concentrations above 2 µM, addition of LiCl alone had a strong effect on phosphorylation (~2-fold increase, Fig. 9.2C+D). When LiCl was added together with CKI7, phosphorylation and translocation were indistinguishable from cells treated with CKI7 alone (Fig. 9.2C+D). Therefore, GSK-3 inhibition seemed to have a weak effect at low ionomycin concentration and a stronger effect at high concentrations; for CK1 inhibition, this picture was reversed.

In summary, several data sets were collected, where phosphorylation and localization were measured simultaneously. It was found that import and export followed the dynamics of de- and rephosphorylation with a delay, as expected, and that the deactivation process proceeded with slower kinetics than activation. In addition, it was analyzed, how the kinase inhibitors affected NFAT kinetics and the ionomycin dose-response curve. Comparison of the inhibitors' effects at different stimulus strengths yielded rather unexpected results, showing that inhibitor effects depended on the ionomycin concentration. To analyze, which processes were responsible for the observed effects, in the next step, a mathematical model was developed based on the presented data set.

Chapter 10

Development of a mathematical model

In this chapter, the development of a quantitative mathematical model of the signaling network controlling NFAT activity is described. In section 10.1, a mathematical description is derived of the reactions that mediate NFAT de- and rephosphorylation and nuclear transport. The resulting model describes a fixed network structure with a large number of kinetic parameters. In section 10.2, it is verified that this model structure can in principle account for the experimental data. In section 10.3, the model is simplified through equalizing related parameters. In a step-up model selection strategy additional complexity is added in a step-wise manner until no further significant improvement in fitting the data is possible. In section 10.4 and 10.5, confidence intervals for the parameters are estimated and the parameter values are analyzed.

10.1 Model description

In the data set described in the previous chapter, localization of NFAT and phosphorylation of the SP2/3 motifs was quantified in combination with perturbation of the kinase activities that target the SRR1 and SP2 motifs. To describe this data set, the SP2(/3) and SRR1 motifs must be distinguished in the model (schematic representation in Fig. 10.1). Although each motif contains several phosphorylation sites, it was assumed for simplification that they were targeted in a single step. To describe the phosphorylation states of the two motifs, a two-letter notation is used, where the first letter indicates the status of the the SRR1 motif that is targeted by CK1, and the second letter stands for the SP2 motif that is phosphorylated by GSK-3 (e.g. OP means that SRR1 is dephosphorylated and SP2 is phosphorylated) (Fig. 10.1). Consequently, the model distinguishes four different phosphory-

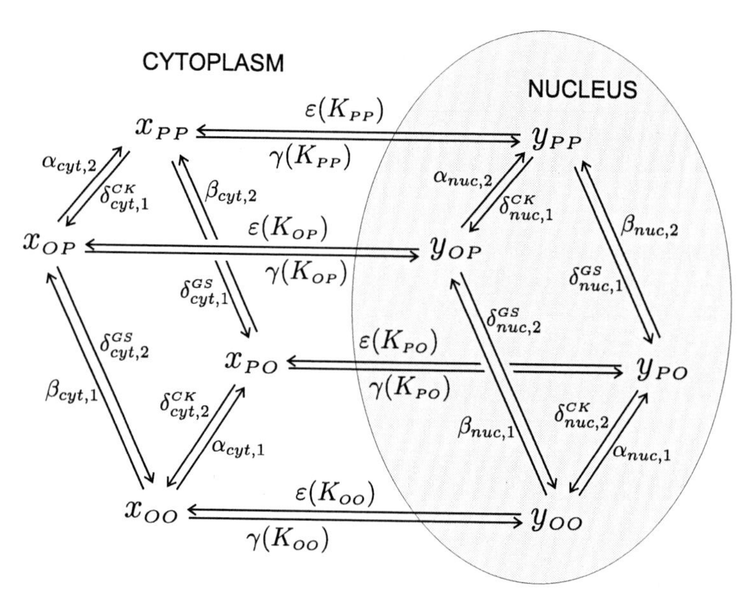

Figure 10.1. Model Scheme. NFAT can adopt four different phosphorylation states (PP, OP, PO, OO), where the first letter represents the SRR1 motif, while the second letter stands for the SP2 motif. 'P' denotes the phosphorylated state of the motif and 'O' the dephosphorylated state. In each phosphorylation state, NFAT can reside in the nucleus (y) or in the cytoplasm (x). The arrows represent the state transitions and are labeled with the symbol of the rate constant for the respective reaction.

Table 10.1. Model Equations

$$\begin{aligned}
\dot{x}_{PP} &= \alpha_{cyt,2} \cdot x_{OP} + \beta_{cyt,2} \cdot x_{PO} - (\delta^{CK}_{cyt,1} + \delta^{GS}_{cyt,1}) \cdot x_{PP} + \varepsilon_{PP} \cdot y_{PP} - \gamma_{PP} \cdot x_{PP} \\
\dot{y}_{PP} &= \alpha_{nuc,2} \cdot y_{OP} + \beta_{nuc,2} \cdot y_{PO} - (\delta^{CK}_{nuc,1} + \delta^{GS}_{nuc,1}) \cdot x_{PP} - \rho \cdot \varepsilon_{PP} \cdot y_{PP} + \rho \cdot \gamma_{PP} \cdot x_{PP} \\
\dot{x}_{OP} &= -(\alpha_{cyt,2} + \delta^{GS}_{cyt,2}) \cdot x_{OP} + \delta^{CK}_{cyt,1} \cdot x_{PP} + \beta_{cyt,1} \cdot x_{OO} + \varepsilon_{OP} \cdot y_{OP} - \gamma_{OP} \cdot x_{OP} \\
\dot{y}_{OP} &= -(\alpha_{nuc,2} + \delta^{GS}_{nuc,2}) \cdot y_{OP} + \delta^{CK}_{nuc,1} \cdot y_{PP} + \beta_{nuc,1} \cdot y_{OO} - \rho \cdot \varepsilon_{OP} \cdot y_{OP} + \rho \cdot \gamma_{OP} \cdot x_{OP} \\
\dot{x}_{PO} &= -(\beta_{cyt,2} + \delta^{CK}_{cyt,2}) \cdot x_{PO} + \delta^{GS}_{cyt,1} \cdot x_{PP} + \alpha_{cyt,1} \cdot x_{OO} + \varepsilon_{PO} \cdot y_{PO} - \gamma_{PO} \cdot x_{PO} \\
\dot{y}_{PO} &= -(\beta_{nuc,2} + \delta^{CK}_{nuc,2}) \cdot y_{PO} + \delta^{GS}_{nuc,1} \cdot y_{PP} + \alpha_{nuc,1} \cdot y_{OO} - \rho \cdot \varepsilon_{PO} \cdot y_{PO} + \rho \cdot \gamma_{PO} \cdot x_{PO} \\
\dot{x}_{OO} &= -(\beta_{cyt,1} + \alpha_{cyt,1}) \cdot x_{OO} + \delta^{GS}_{cyt,2} \cdot x_{OP} + \delta^{CK}_{cyt,2} \cdot x_{PO} + \varepsilon_{OO} \cdot y_{OO} - \gamma_{OO} \cdot x_{OO} \\
\dot{y}_{OO} &= -(\beta_{nuc,1} + \alpha_{nuc,1}) \cdot y_{OO} + \delta^{GS}_{nuc,2} \cdot y_{OP} + \delta^{CK}_{nuc,2} \cdot y_{PO}) - \rho \cdot \varepsilon_{OO} \cdot y_{OO} + \rho \cdot \gamma_{OO} \cdot x_{OO}
\end{aligned}$$

$$\varepsilon_{XX} = \frac{\varepsilon_0}{1 + K_{XX}} \qquad \gamma_{XX} = \frac{\gamma_0 \cdot K_{XX}}{1 + K_{XX}}$$

lation states, the phosphorylated (OO), the dephosphorylated (PP) and two partially phosphorylated forms (PO, OP).

The enzymes that catalyze de- and rephosphorylation are not explicitly included in the model. Instead a rate constant is assigned to each reaction that incorporates enzyme concentration and catalytic activity for the respective substrate. The rate constants describing the phosphatase calcineurin are symbolized by δ, activity of CK1 by α and GSK-3 by β. The model parameters are summarized in Table 10.2). Enzyme activities might differ between nucleus and cytoplasm due to differences in enzyme concentrations and de- and rephosphorylation rates of one motif might depend on the phosphorylation state of the other motif. Each phosphorylation state can by imported into the nucleus (rate constants γ) and exported from the nucleus (rate constants ε). All states can be localized in the nucleus (denoted as x) or in the cytoplasm (y). The network in Fig. 10.1 was translated into a mathematical model, using ordinary differential equations. Linear mass action kinetics were used to describe all reactions, assuming that the rate of each reaction is proportional to the substrate concentration (Table 10.1).

Since NFAT localization is controlled by its phosphorylation status, import (γ) and export rates (ε) differ between the different states. This observation lead to the hypothesis that NFAT exists in an equilibrium between two conformations, exposing either the NLS or the NES, and that the fraction residing in the import and export conformation, respectively, was determined by the phosphorylation state (Okamura et al., 2000). In the model, the conformational equilibrium is described by the equilibrium constants

$$K = \frac{[\text{Import conformation}]}{[\text{Export conformation}]} \tag{10.1}$$

The import and export conformations are transported with the rate constants γ_0 and ε_0, respectively. Assuming a fast conformational equilibrium, the phospho-specific transport rates can be described as functions of the respective equilibrium constants K (Table 10.1, bottom). A detailed derivation of this relationship is given in Appendix B.3. Because NFAT moves between different compartments, the model takes into account the differing volumes of nucleus and cytoplasm (detailed description in Appendix B.3). The ratio of cytoplasmic to nuclear volume ρ was set to $\sim$3 (Uwe Vinkemeier, personal communication).

10.2 Model fitting

In the next step, it was tested whether the model derived in the previous section could in principle account for the experimental data shown in Fig. 9.2.

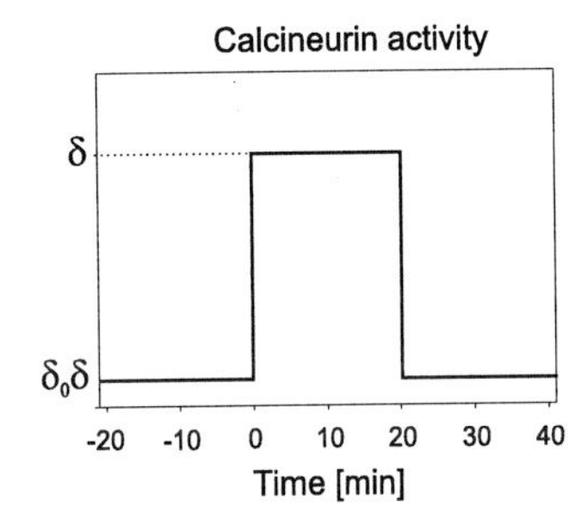

Figure 10.2. Kinetics of calcineurin activity. Upon ionomycin stimulation calcineurin activity is increased to δ. In the absence of stimulation and after addition of CsA, a low basal phosphatase activity was assumed ($\delta_0 \cdot \delta$).

The kinase activities (α and β) were assumed to be constant throughout the experiment. The assumed temporal profile of the phosphatase activity is shown in Fig. 10.2. Its rate constant upon addition of 2 µM ionomycin is denoted as δ; the low basal activity prior to stimulation and upon CsA addition is described as a fraction δ_0 of δ. The effect of the kinase inhibitors is described by the parameter ω, which denotes the fraction of kinase activity (α, β) that remains in the presence of the inhibitor. The efficiency of the kinase inhibitors ω are not necessarily identical. The simulation of dose-response experiments is described in detail in the Appendix B.4. In short, a sigmoidal (Hill-type) relationship between the ionomycin concentration and the resulting phosphatase activity was assumed, described by two parameters (κ and the Hill coefficient λ).

Assuming that all 27 model parameters were independent from each other, the model was fitted to the data shown in Fig. 9.2. A detailed description of the fitting procedure can be found in Appendix B.5. It was shown that the model structure was able to account for the experimental observations (Fig. B.3 in Appendix B.5).

10.3 Model selection

Having found that the model structure was able to explain the experimental measurements, the minimal model complexity required to reproduce the data was determined. The applied model selection strategy is explained in detail in Appendix B.3 and will only be shortly outlined here. First, the complexity of the full model was radically reduced from 27 to 10 independent parameters, by assuming for example that all phosphorylation reactions would proceed with the same rate constants irrespective of which motif was targeted and of where the reaction took place. Since the simplest model was not able to account for the data, several additional assumptions were tested, for example that the reaction rates differed between nucleus and cytoplasm. Models of distinct complexities were compared with statistical methods, using the weighed sum of squared residuals (χ^2) to evaluate the goodness of the fit. Then, the assumption that would improve the fit most, was selected. This procedure was repeated until no further significant improvement was possible.

The final model found through the model selection procedure contained 16 free parameters and is shown schematically in Fig. 10.3A. The following three assumptions were found to be crucial to simulate the data:

1. The rate constants of the phosphorylation reactions (α, β) depend on the phosphorylation state of the **not-targeted** motif (e.g. $\alpha_1 \neq \alpha_2$).

2. LiCl and CKI7 exhibit different inhibition efficiencies ($\omega_{CK} \neq \omega_{GS}$).

3. The transport rates of the OO and OP states are different ($K_{OO} \neq K_{OP}$), while the transport rates of the OO and PO can be set equal ($K_{OO} = K_{PO} \neq K_{XO}$).

Moreover, the model selection procedure showed that the kinase activities can be assumed to be equal in the nucleus and in the cytoplasm. The phosphatase activity is much higher in the cytoplasm, which has also been shown previously (Kilka et al., 2009; Liu et al., 1991). Assuming a non-zero phosphatase activity also in the nucleus resulted in a small, but significant improvement of the fit.

The final model can explain most features of experimental measurements (Fig. 10.3B-E). Similar to our experimental observations, also in the simulation activation proceeded with faster kinetics than deactivation and translocation followed phosphorylation with a delay (Fig. 10.3B+C). Moreover, the different effects of the kinase inhibitors on the resting state were reproduced quite well by the model, in that LiCl had virtually no effect, while CKI7 lead to a significant increase in dephosphorylation and translocation (Fig. 10.3B+C compare green and red). The model also reproduced the finding that the inhibitors had additive effects in response to 2 µM ionomycin, while at lower concentrations LiCl alone had a weak effect that was much stronger in the presence of CKI7 (Fig. 10.3D+E). The main disagreement between model and simulation was observed in the import kinetics. In the experiment, the nuclear import proceeded with significantly faster kinetics than in the simulation (Fig. 10.3C). Since also the complex model with 27 parameters could not explain the fast import kinetics (Fig. B.3), they might be controlled by processes not described in the model structure. Moreover, the strong effect of LiCl on the dephosphorylated fraction at high ionomycin concentrations was not completely reproduced by the model (Fig. 10.3D).

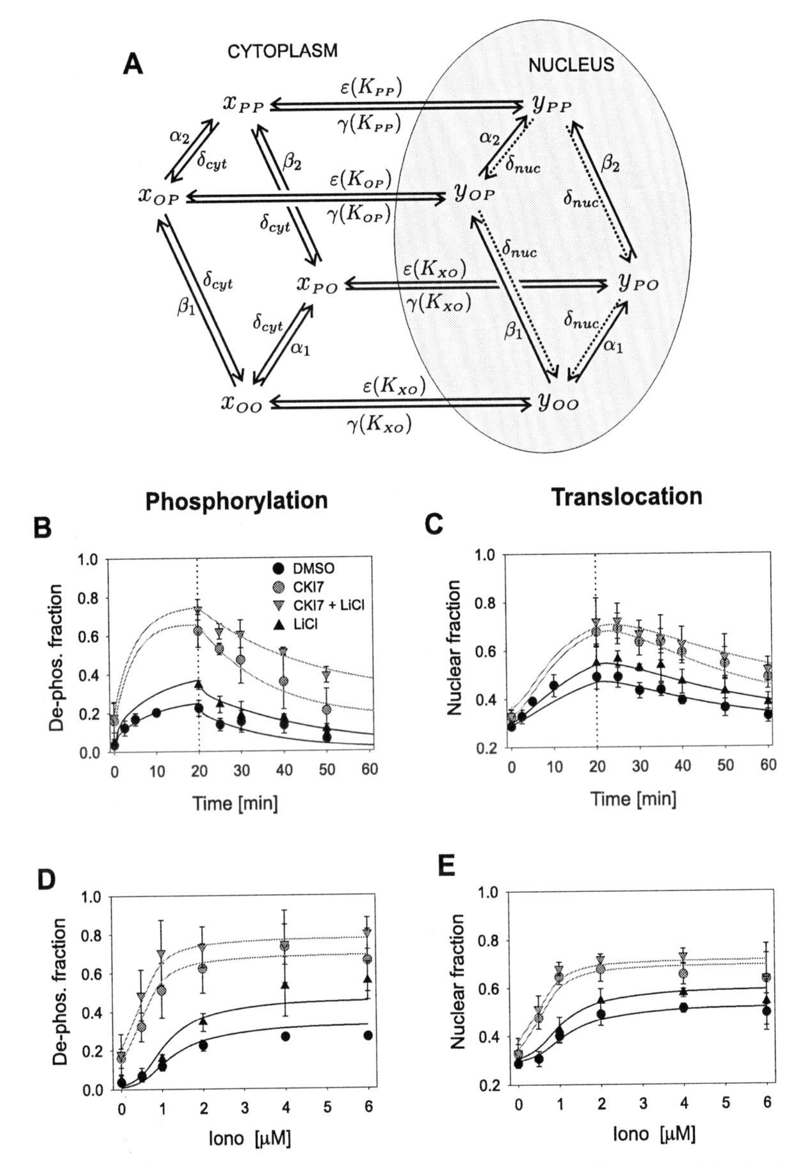

Figure 10.3. Best fit of the experimental data. **(A)** Structure of the final model obtained through the model selection procedure described in Appendix B.6. **(B-E)** The 16 model parameters were fitted to the experimental data described in Fig. 9.2. Lines represent the model simulation and symbols the experimental data.

Table 10.2. Model parameters

Symbol	Description	Value	90% Conf. int.
α_1	phosphorylation of the OO-state by CK1	0.010 min^{-1}	[0 ; 0.030]
α_2	phosphorylation of the OP-state by CK1	9.9 min^{-1}	[2.0 ; 10]
β_1	phosphorylation of the OO-state by GSK-3	0.046 min^{-1}	[0.032 ; 0.056]
β_2	phosphorylation of the PO-state by GSK-3	2.2 min^{-1}	[1.1 ; 7.1]
δ_{cyt}	cytoplasmic dephosphorylation at 2μM Iono	0.27 min^{-1}	[0.17 ; 0.42]
δ_{nuc}	nuclear dephosphorylation at 2μM Iono	0.032 min^{-1}	[0.006 ; 0.077]
δ_0	fraction of phosphatase activity in resting cells	5.5 %	[3.4 ; 9.4]
K_{PP}	equilibrium constant between the import and the export conformation of the PP state	0.010	[0.001 ; 0.092]
K_{OP}	equilibrium constant of the OP state	0.006	[0 ; 0.058]
K_{XO}	equilibrium constant of the OO and PO states	0.095	[0.009 ; 1]
γ_0	nuclear import	1.5 min^{-1}	[0.2 ; inf]
ε_0	nuclear export	0.013 min^{-1}	[0.009 ; 0.016]
ω_{CK}	inhibition of CK1 by CKI7	0.01 %	[0 ; 0.8]
ω_{GS}	inhibition of Gsk3 by LiCl	53 %	[46 ; 61]
κ	Ionomycin dependence of calcineurin activity	1.4	[1.0 ; 1.8]
λ	Ionomycin dependence of calcineurin (hill coef.)	2.0	[1.4 ; 3.0]

10.4 Parameter estimation

This project aimed at estimating the rate constants of the reactions that control NFAT activity. In the previous section, the model was fitted to experimental data and the estimated parameter values are shown in Table 10.2. To draw reliable conclusions from the values, it must be analyzed, how well a parameter can be estimated from the data. Therefore it was tested, how much the parameter value can be changed until the goodness of the fit decreases significantly. In Appendix B.7 it is described in detail, how confidence intervals for the parameter values were derived (Raue et al., 2009). Confidence intervals for the model are given in Table 10.2).

One of the questions that were addressed in this project concern the mode of phosphorylation and dephosphorylation. Do these reactions proceed in a strict order (e.g. first SRR1 then SP2 or vice versa), suggesting a sequential mechanism? Or can each motif be targeted with the same efficiency, which would propose a random mechanism? Is there cooperativity in these reactions, where the phosphorylation state of one motif might influence the modification rate of the other one? These questions can be answered through analyzing the rate constants estimated for the phosphorylation and dephosphorylation reactions: If for example, phosphorylation of the SRR1 motif (PO) from the dephosphorylated state (OO) would proceed with a faster rate than modification of SP2 (OP), SRR1 would be preferentially phosphorylated, resulting in a sequential mechanism (OO $\rightarrow$PO $\rightarrow$PP). Cooperativity would occur, if the reaction rates for modification of one motif depended on the phosphorylation state of the other motif.

The phosphatase activity in the cytoplasm was found to be identical for all steps (Table 10.2, δ_{cyt}). Therefore the dephosphorylation reaction seemed to proceed in a random manner with no cooperative effects. By contrast, a strong cooperativity was found in the phosphorylation reactions: The first phosphorylation step of the completely dephosphorylated state occurred with very low rates (α_1,β_1<0.1 min^{-1}, Table 10.2), while the second phosphorylation step proceeded with >10-fold faster kinetics (α_2,β_2>1 min^{-1}, Table 10.2). Therefore, phosphorylation of one motif strongly facilitated modification of the other one in the model. To understand, whether phosphorylation proceeds with a random or a sequential mode, CK1 and GSK-3 activities must be compared. Although the estimated rate constants of CK1 and GSK-3 differed up to five-fold in the best fit, their overlapping confidence intervals suggested that their values were not well defined based on the experimental data used. Therefore, the presented analysis is compatible with a random or a slightly sequential mode of phosphorylation.

The second question that must be answered to understand the roles of the kinases in controlling NFAT activity, concerns the transport properties of the

Table 10.3. Equilibrium constants. For fixed import rate γ_0=0.5 min^{-1}, the equilibrium constant and their confidence intervals are given.

Parameter	Value	90% Conf. int.
K_{PP}	0.030	[0.022 ; 0.048]
K_{OP}	0.013	[0 ; 0.035]
K_{OO}	0.30	[0.19 ; 0.41]
K_{PO}	non-identifiable	

partially phosphorylated states that are given by their equilibrium constants K. Large confidence intervals were found for the equilibrium constants (Table 10.2). Therefore further analyses had to be performed, described in Appendix B.7. It was found that the relation of the equilibrium constants to each other was well defined, but their absolute values could not be estimated. K_{OO} was found to be ~10-fold higher than K_{PP} and K_{OP} (Fig. B.6 in Appendix B.7). Only K_{PO} was truly non-identifiable. If the import rate constant γ_0 was set to a realistic, rather low value, such as 0.5 min^{-1}, the other three equilibrium constants could be identified (Table 10.3). For K_{PP} and K_{OP} (~0.01) similar values were found, and both were significantly smaller than K_{OO} (~0.3). Therefore, we can say that K_{OP} is similar to K_{PP} and about 10-fold smaller than K_{OO}, but the properties of the PO state still remain to be determined.

In summary, several surprising results were found from the parameter values. Analysis of the equilibrium constants suggested that, in terms of transport rates, the dephosphorylated state (OO) resembled the partially phosphorylated PO state and that the fully phosphorylated state (PP) was similar to the OP state. Therefore, the transport properties seemed to be mainly controlled by the SP2 motif (because $K_{XO} \gg K_{XP}$), which is contrary to what has been proposed previously (Okamura et al., 2000) and which will be discussed in more detail in chapter 11.

If we compare the import and export rate constants, γ_0 and ε_0, it is surprising that the import is estimated to be two orders of magnitude faster than the export. It is typically found that export proceeds with somewhat slower kinetics than import, but such a large difference is unrealistic and suggests that processes not described in the model might delay the export, such as retention through specific nuclear proteins.

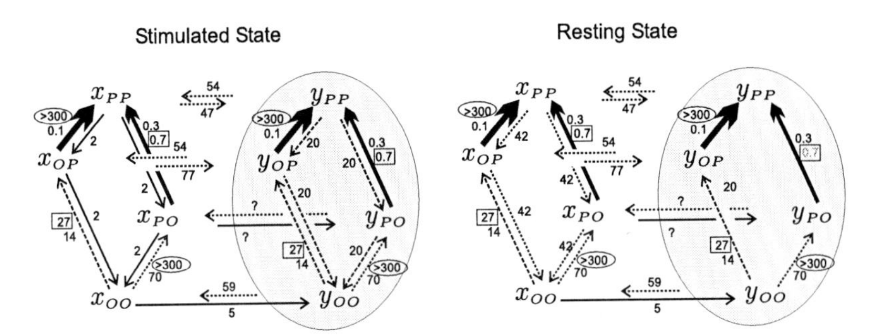

Figure 10.4. Pathway half-times in the NFAT reaction network. The parameter values in Table 10.2 were used to calculate the half time for each reaction. The numbers indicate the half times in the absence of inhibitors in minutes, for the stimulated (left) and for the resting state (right). In circles and squares, the values in the presence of CKI7 and LiCl, respectively, are given. Thicker and solid arrows indicate faster reactions, while dashed and dotted lines stand for slower reactions. The half times of the processes marked with question marks could not be identified from the data.

10.5 Model Analysis

One of the main objectives of this project, was to identify the roles played by the different kinases. To address this questions, the properties of the partially phosphorylated states and the modes of de- and rephosphorylation were investigated in the previous section. The results are summarized in Fig. 10.4, where the half times of all processes in the model are indicated. As already discussed in the previous section, the transport properties of the partially phosphorylated OP state, where only the SP2 motif is phosphorylated, resemble more the fully phosphorylated (PP) than the fully dephosphorylated molecule (OO), in that it is slowly imported and exported, with half-times around 1 hr. By contrast, the fully dephosphorylated state is imported within 5 min. The transport properties of the PO state, where only the SRR1 motif is phosphorylated, could not be identified from the data. In summary, it can be concluded that the OP state cannot be imported efficiently.

Surprisingly, the partially phosphorylated states seem to be a much better substrate for the kinases ($t_{\frac{1}{2}} < 1$ min) than completely dephosphorylated NFAT ($t_{\frac{1}{2}} > 10$ min), suggesting a strong cooperativity in the phosphorylation reaction. By contrast, for the dephosphorylation reaction, no cooperativity was found; all steps proceeded with rather fast kinetics ($t_{\frac{1}{2}} = 2$ min). The net dephosphorylation, however, depends on the balance of phosphatase and kinase activity. As soon as one motif is de-phosphorylated, the other mo-

Table 10.4. Normalized control coefficients

Enzyme/Process	$NFAT_{nuc}$ (Rest.)	$NFAT_{nuc}$ (Stim.)	Import	Export	$NFAT^{XO}_{nuc}$
Calcineurin(δ)	0.04	0.41	-0.08	0.09	1.38
CK1(α)	0	-0.03	-0.02	-0.10	-0.22
Gsk3(β)	-0.04	-0.24	0.02	-0.27	-0.93
import (γ)	0.70	0.52	-0.21	-0.43	0.20
export (ε)	-0.68	-0.41	-0.08	-0.42	-0.13

tif follows rapidly, because the first step is rate-limiting due to the strong counteracting kinase activity. Therefore, the model predicts that the partially phosphorylated states are only present in very small amounts in the cell. Interestingly, both phosphorylation pathways (with OP and PO intermediates) seem to exhibit strong cooperativity and no preferred reaction sequence could be identified. Therefore, the model parameters are consistent with a random phosphorylation mechanism and there is no clear evidence for sequential phosphorylation or dephosphorylation.

To further analyze the roles played by CK1 and GSK-3, a control analysis was performed. Control coefficients indicate, how much a parameter change affects the response of the system. In Table 10.4, normalized control coefficients are given for the reaction rates of calcineurin, CK1 and GSK-3 and for the import and export rate constants. To understand, which processes regulate the amount of NFAT in the nucleus, control coefficients for the nuclear fraction in resting and in stimulated cells were analyzed (Table 10.4, 1.+2. column). In the resting state, only the transport rates exerted strong control (+/- 0.7), while the enzymes controlling the phosphorylation state had virtually no effect (cc<0.1). Because NFAT is ~100-fold faster rephosphorylated than dephosphorylated, it remains stably phosphorylated in the cytoplasm and is not affected by a small change of these rates. Only a strong perturbation of the kinase activity, such as addition of CKI7, can change NFAT's localization significantly, as seen in the presented experiments. In the stimulated state, a different picture emerges. Here, not only the transport rate constants exert significant control, but also calcineurin (cc=0.41) and GSK-3 (cc=-0.24) affect the nuclear fraction.

In the next step, control coefficients for the import and export kinetics during activation and deactivation were analyzed (Table 10.4, 3.+4. column). While the de- and rephosphorylation reactions did not affect the import kinetics, export kinetics did respond to changes of GSK-3 activity (cc=-0.27). This finding was independent of the assumed transport properties of the PO state (K_{PO}). Instead it can be attributed to the fact that rephosphorylation of the SP2 motif is faster than rephosphorylation of the SRR1 motif, making it the main rephosphorylation pathway (via the OP state) (Fig. 10.4).

In summary, in the presented model, GSK-3 can be viewed as an export kinase (Beals et al., 1997b), but, in contrast to previously proposed models, CK1 does not act as a maintenance kinase (Okamura et al., 2004). The proposed role of CK1 as maintenance kinase is largely based on the observation that inhibition of CK1, but not inhibition of GSK-3 results in partial nuclear localization in the resting state (Fig. 9.2). In the model, however, this observation is explained by the different efficiencies of the CK1 and GSK-3 inhibitors. Similar inhibition of CK1 and GSK-3 should result in comparable nuclear accumulation, because the first dephosphorylation step is rate-limiting. Once the partially phosphorylated state is reached, complete dephosphorylation occurs rapidly, because the counter-acting kinase activity is low. This mechanism also explains, why LiCl has no detectable effect in the resting state, but results in an increase of the nuclear and de-phosphorylated fractions in the stimulated state or in combination with CKI7. In both scenarios, NFAT reaches the OP state easily, whose further dephosphorylation can be affected by GSK-3 inhibition, even when the inhibitor is inefficient.

In the last step, it was analyzed how the transcriptionally active form of NFAT1, namely SP2 dephosphorylated NFAT (PO,OO) in the nucleus, is controlled (Okamura et al., 2000; Neal and Clipstone, 2001). The strongest positive control was exerted by calcineurin (cc=1.38) and the strongest negative control be GSK-3 (-0.93). These are the highest control coefficient found in this analysis and suggest that the system is designed in a way that gives calcineurin and GSK-3, both of which are regulated by TCR stimulation, maximal control over NFAT's transcriptional activity.

Chapter 11

Discussion

In the presented study, a mathematical model of NFAT activation, including phosphorylation and nuclear transport, was developed based on experimental data. Analysis of the model parameters revealed strong cooperativity in the phosphorylation reaction: As soon as one motif was phosphorylated, modification of the other motif was strongly facilitated. Regarding the properties of the partially phosphorylated states, the model revealed that dephosphorylation of the SRR1 motif was insufficient to induce import, in contrast to previously published observations (Okamura et al., 2000). The transport properties of the other partially phosphorylated form, where only the SP2 motif is dephosphorylated, could not be estimated from the experimental data used.

11.1 Cooperativity in NFAT de- and rephosphorylation

The dephosphorylation of NFAT has been proposed to proceed in a sequential fashion (Okamura et al., 2000; Sheridan et al., 2002). The presented analysis, however, cannot support this hypothesis. Originally, the idea arose from two experimental observations: It was found that at low ionomycin concentrations, SRR1 was preferentially dephosphorylated and that a serine-to-alanine mutation of the SRR1 motif reduced the stimulus strength required for dephosphorylation of the SP2 motif[1] (Okamura et al., 2000). From the former observation it was concluded that calcineurin could access the SRR1 motif, which is adjacent to a calcineurin docking site, more easily than the other phosphorylation motifs. In the presented analysis, however, no support was found for the idea that calcineurin would preferentially target a specific mo-

[1] Substitution of a serine residue by an alanine prevents phosphorylation of that residue. Therefore the mutant mimics NFAT with a constitutively dephosphorylated SRR1 motif.

tif. In a model selection procedure, distinct features, such as motif-specific phosphatase activity, were added and it was tested, whether the addition would improve the goodness of the fit to the experimental data. While motif-specific phosphatase activity did not improve the fit, differential kinase activity was selected as an important feature of the model. Since the two motifs are targeted by different kinases, it is reasonable that their phosphorylation proceeds with distinct rates. Since the kinase activity, counteracting SRR1 dephosphorylation was estimated to be faster than SP2 phosphorylation, a higher fraction of SP2 dephosphorylated NFAT was observed in model simulations (Fig. B.4 in the Appendix). The estimated parameters, however, are associated with too large confidence intervals ($\alpha_2 = [2; 10]$ and $\beta_2 = [1; 7]$) to decide, which motif might be phosphorylated with a higher rate. The tendency found in the best fit, where SRR1-phosphorylation is faster than SP2 phosphorylation, would suggest a preferential dephosphorylation of SP2 at low stimulus strength.

In the second experiment that had suggested sequential dephosphorylation, a mutated form of NFAT was analyzed, where the serine residues in the SRR1 motif were substituted by alanines. The mutant protein was found to be dephosphorylated at the SP2/3 motifs at ~10-fold lower ionomycin concentration than the wildtype protein (Okamura et al., 2000). The data presented in the present study support this observation, showing that inhibition of CK1, which otherwise keeps SRR1 phosphorylated, resulted in a strong increase in the dephosphorylated fraction (which represents the SP2/3 motifs), in particular at low ionomycin concentrations. The presented mathematical model explains this finding with a strong cooperativity in the phosphorylation reactions. In the model selection procedure it was found that a partially phosphorylated substrate was phosphorylated more easily than the completely dephosphorylated form. The estimated parameters revealed a >50-fold acceleration of the phosphorylation reaction, when the other motif is already phosphorylated. This finding can be interpreted as a unspecific priming for further phosphorylation, where partial phosphorylation makes the whole regulatory domain more accessible to kinases. However, in the model cooperative phosphorylation does not result in sequential de- or rephosphorylation. Therefore, also serine-to-alanine mutation of the SP2 motif should result in an increased sensitivity for SRR1 dephosphorylation, which would, however, be difficult to measure experimentally.

11.2 The partially phosphorylated states

In addition to the mode of de- and rephosphorylation, also the transport properties of the partially phosphorylated states determine the functional roles of the different NFAT kinases. It has been previously proposed that dephosphorylation of the SRR1 motif (OP) allows nuclear import, while dephosphorylation of the SP2 motif (PO) does not (Okamura et al., 2000). In the presented analysis, however, we find the opposite: The OP state resembles the completely phosphorylated state in its transport properties; for the PO state the transport properties could not be estimated from the data. Originally, it was proposed that the OP state is imported faster than the completely phosphorylated protein, because a NFAT mutant, where the serine residues of the SRR1 motif were substituted by alanines, was partially nuclear in resting cells (Okamura et al., 2000). In the presented model, this behavior can also be explained. The first dephosphorylation step is limiting and therefore a partially phosphorylated protein that is mimicked by the mutations, reaches the completely dephosphorylated state easily and is then imported. Two previously published observations, however, cannot be explained by the model: Also in the presence of the calcineurin inhibitor CsA the mutant protein was partially nuclear, suggesting that import did not require further dephosphorylation. Moreover, mutation of the SP2/3 motifs did not result in a change in localization, as it would be predicted by the model. This discrepancy between the here described results and these published observations might result from the fact that here a NFAT construct was used that only contained the regulatory domain, while the cited study used the full-length protein, where additional effects might occur.

In the parameter estimation procedure the absolute values of the equilibrium constants describing the ratio of molecules that adopted the importable and the exportable conformation, were not identifiable. However, the ratio between the different equilibrium constants was well defined and when the import rate constant (γ_0) was fixed to a reasonable value, the equilibrium constants could be estimated. The values revealed that for all states a significant fraction (>50 %) adopted the export conformation and therefore all states could in principle be exported. However, nuclear export proceeded with slower kinetics than import (Zhu et al., 1998). This was explained by the fact that the export conformation itself was transported with a very slow rate (ε_0 =0.1 min^{-1}). The fact that dephosphorylated NFAT associates stably with the export receptor Crm1 (Okamura et al., 2004; Kehlenbach et al., 1998; Zhu and McKeon, 1999), makes it unrealistic that the transport reaction itself would proceed with such slow kinetics. More likely, another process, not described by the model, such as retention by specific nuclear factors might be rate-limiting for the export reaction.

11.3 Functional roles of calcineurin and NFAT kinases

In agreement with previous reports, similar kinase activities were found in the nucleus and in the cytoplasm, while the phosphatase activity was 10-fold lower in the nucleus than in the cytoplasm (Loh et al., 1996; Liu et al., 1991; Kilka et al., 2009). Moreover, the presented analysis confirmed previous reports that had demonstrated a basal phosphatase activity also in the absence of stimulation (Okamura et al., 2004). As expected from its important functional role, also in our model, calcineurin exerts strong positive control (control coefficient > 1) over the transcriptionally active form of NFAT. The strongest negative control was exerted by GSK-3, explaining also why its activity is down-regulated during antigenic stimulation. Co-stimulation through CD28 results in inhibits phosphorylation of GSK-3, thereby increasing transcriptional activity of NFAT (Cross et al., 1995; Diehn et al., 2002).

Interestingly, the two processes that, according to the presented model, exert the strongest control over NFAT's transcriptional activity have also been implicated in a genetic disease: The pathology of trisomy21 which causes Down syndrome has been partially attributed to deregulated NFAT-signaling (Arron et al., 2006; Baek et al., 2009). An inhibitor of calcineurin (DSCR1) and the NFAT kinase DYRK1A are located on chromosome 21. According to the model, reduced calcineurin activity, which would result from increased doses of DSCR1, and increased DYRK, activity, which is partially represented by GSK-3 in the model, should result in a strong defect in NFAT signaling (Appendix B.7).

GSK-3 has previously been proposed to act as an export kinase (Beals et al., 1997b), which was confirmed in the presented analysis, but no support was found for the proposed function of CK1 as an import and maintenance kinase. In this context, it might also be necessary to include DYRK explicitly in the analysis to fully understand NFAT regulation. Overexpression and knock-down of DYRK have been shown to strongly affect import, export and transcriptional activity (Arron et al., 2006; Gwack et al., 2006). Moreover, it can act as a priming kinase for GSK-3 and CK1, making it an important regulator (Arron et al., 2006; Gwack et al., 2006).

11.4 Outlook

In the future extension of this study it would be insightful to explicitly include DYRK in the analysis. Activity of DYRK kinases can be perturbed by siRNA. It is, however, technically difficult to determine the phosphorylation status of all three phosphorylation motifs, which would be necessary to develop a reliable model. Possibly, this question could be addressed by quantitative phospho-proteomics using mass spectrometry. The presented model, represents a rather simplified picture of reality in that only two phosphorylation motifs are distinguished, which in reality consist of 13 different residues. The potential role of multisite phosphorylation in NFAT regulation has been addressed by mathematical models in the past (Salazar and Höfer, 2003). If multisite phosphorylation is controlled by a cooperative mechanism, it can result in an ultrasensitive, highly cooperative dose-response relationship. In the presented dose-response measurements a weak sigmoidality was found as expected from ultrasensitivity. It is, however, not possible to determine, whether it arises from multiple dephosphorylation or from cooperativity in calcineurin activation. The latter has been proposed to arise from the fact that several Ca^{2+} ions bind to Calmodulin, resulting in a hill coefficient of $\sim$3 (Stemmer and Klee, 1994). Similarly, in the presented model a hill coefficient (λ) of 2 was found.

As already discussed in the previous sections, the presented model failed to reproduce a few experimental observations, related to the role of CK1. In the unbiased model selection procedure applied here, the fact that CK1 inhibition had a stronger effect than GSK-3 inhibition, was attributed to differential efficiency of the respective inhibitors. Other observations, however, attribute this finding to differential transport properties of the partially phosphorylated states. Therefore it would be worthwhile to analyze, whether the presented data could be explained with assuming a similar efficiency of the two inhibitors, as suggested by *in vitro* experiments (Okamura et al., 2004). In summary, several open questions remain to be addressed in the future and extended models might have to be developed.

Part III

Concluding remarks

In this thesis, two different questions in molecular immunology were tackled through an approach, where mathematical modeling was tightly linked to quantitative experiments. It is becoming more and more obvious that probably all processes in mammalian organisms are regulated by rather complex regulatory networks. To extend our understanding of the underlying regulatory principles beyond a purely descriptive level, such a quantitative approach is strictly required. Here kinetic models are a useful tool to rationalize complex data sets, predict the systems behavior to perturbations and to understand the system properties on a quantitative level.

In the approach used in the present study the right level of simplification is crucial to analyze a complex network. A reliable model can only be developed, when the number species and processes that can be measured experimentally is comparable to the number of variables and reactions described by the model. Moreover, one should be able to experimentally perturb a similar number of processes. In the project on the Th1 gene network described in part I of this these, these conditions were met in that a network of three genes turned out to be sufficient to account for a large body of experimental data. Moreover, several processes could be perturbed rather easily, because they were mediated by extracellular signals that can be manipulated experimentally. The main shortcoming of the experimental data produced was the rather high inter-experiment variability. Such a behavior is not unusual, when primary cells are used. Therefore a compromise must be found regarding the data quality and the extent to which it represents the situation *in vivo*. However, the project aimed at elucidating the *structure* of the underlying network and it was not necessary to accurately estimate its kinetic parameters. For this purpose the data quality turned out to be sufficient.

In the second part of this thesis, the regulatory network controlling NFAT activity was analyzed. Here, the model described eight species that differed in their localization (nuclear or cytoplasmic) and in their phosphorylation state. It was, however, only possible to measure localization and the state of one out of two phosphorylation motifs. Therefore the number of model variables exceeded the number of measurable species, probably limiting the predictivity of the resulting model. Moreover, a third phosphorylation motif was not even included in the model, but is known to be involved in the regulation of NFAT. Since so far the technical possibilities for determining the phosphorylation state of NFAT are rather limited, mutant proteins with a reduced number of phosphorylation motifs could be used to solve this problem. Their behavior, however, would represent the situation *in vivo* even less.

A third important criterion for developing a meaningful model is the incorporation of data that links the analyzed processes to the physiological function of the network. For the Th1 gene network it was very insightful that for all kinetic measurements during the differentiation process also the outcome of this process was quantified. Through correlating expression of the Th1-specific transcription factor T-bet with the differentiation efficiency, it was found that T-bet must be expressed during a specific time window in order to drive differentiation.

In the project described in the second part, the link to the network's physiologic function was somewhat missing. A NFAT construct was used for the experimental measurements that cannot bind to DNA. While this limits the "side effects" of the construct (e.g. cytotoxicity[2]) and reduces the number of processes that must be included in the model (i.e. DNA binding can be neglected), it does not allow us to link the measurements to target gene transcription. This case illustrates nicely the trade-off between simplifying the investigated system to a manageable number of species and processes and the possibility to link the results to the physiological function of the system.

Overall, the described approach turned out to be very useful for answering a number of important questions regarding the regulatory networks active in T-helper cells. In future studies, the developed models will hopefully be extended by including additional players. The possibilities of measuring many species simultaneously are ever increasing in the current "-omics" era. However, the modeling tools used here can only be applied in a meaningful way, when the system's size is somewhat restricted. Therefore, the challenge of how to approach the true biological complexity remains open.

[2]It was actually attempted to compare the used construct to a second one that would be able to bind to DNA. However, it was not possible to obtain stable transfectants, because the construct seemed to exert a cytotoxic effect, possibly through interfering with endogenous transcription factors in a dominant-negative manner.

Part IV

Materials and Methods

Chapter 12

Experimental techniques

12.1 Mice, antibodies, chemicals

Balb/c, C57/Bl6 and *Ifngr*$^{-/-}$ on C57/Bl6 background were bred under pathogen free conditions. Antibodies to IFN-γ (AN18.17.24), IL-12 (C17.8), IL-4 (11B11), CD3 (145-2C11), CD28 (37.51), CD4 (GK1.5), CD62L (MEL14) and CD44 (IM7) were prepared from hybridoma supernatants. Phorbol mysterate acetate (PMA), LiCl, ionomycin and Brefeldin A were supplied by Sigma-Aldrich and CKI7 by Seikagaku. The calcineurin inhibitor NCI3 was a kind gift of M. Sieber (Sieber et al., 2007). Cyclosporine A (CsA) was obtained from AWD.

12.2 T-cell culture

Purification of naive Th-cells

Naive $CD4^+$ T-cells were purified by magnetic sorting. Fresh splenocytes from 6-10 week old mice were stained with a Fluorescein-Isothiocyanate-conjugated CD4-specific antibody and CD4 expressing cells were purified using the FITC-Multisort-Kit (Miltenyi Biotec). Subsequently, naive T-cells that express high levels of CD62L were purified using CD62L-Microbeads (Miltenyi Biotec). A purity of 98-99 % was reached. For experiments that required a purer population of naive T-cells, splenocytes were stained with antibodies to CD44 (FITC), to CD62L (PE) and to CD4 (Cy5) and $CD44^{low}CD62L^{hi}CD4^+$ cells were purified by cytometric sorting (100 % purity).

T-cell differentiation

Naive cells ($2 \cdot 10^6$ cells/ml) were stimulated with platebound CD3-specific antibodies (10 g/ml) for 48 hrs. Soluble CD28-specific antibodies (1 g/ml), IL-2 (10 ng/ml) and blocking antibodies to IL-4 (10 g/ml) were included in the medium. To induce Th1 differentiation IL-12 (5 ng/ml) was added to the culture. Where indicated, IFN-γ (10 ng/ml), Cyclosporine A (50 nM) or blocking antibodies to IFN-γ (10 g/ml) or IL-12 (10 g/ml) were used. Cells were cultured in RPMI 1640 medium (Invitrogen) supplemented with 100 units/ml penicillin, 0.1 mg/ml streptomycin (Invitrogen), 10 % heat-inactivated fetal calf serum, and 10 μM β-mercaptoethanol,

12.3 Intracellular flow cytometry

With intracellular flow cytometry the amount of a certain protein per cell can be quantified on the single-cell level. First the cells are fixed, then their membrane is permeabilized. The protein of interest is stained with fluorescently labeled antibodies which can be detected in the flow cytometer. This technique is well established for assessing cytokines expression, but is also increasingly used to detect transcription factors.

Intracellular cytokine staining

After six days of culture, dead cells were removed by gradient centrifugation using Histopaque-1083 (Sigma-Aldrich). To mimic secondary antigenic stimulation, the cells were incubated with 10 ng/ml PMA and 1 g/ml ionomycin in a concentration of $2 \cdot 10^6$ cells/ml. After 1 hr Brefeldin A (5 g/ml) was added to block secretion and after another 3 hrs cells were fixed for 15 min in 2 % formaldehyde. Cells were permeabilized with 0.5 % saponin and stained intracellularly with antibodies to IFN-γ (PE), to IL-4 (Alexa488, BD) and to CD4 (Cy5) and measured by flow cytometry using a FACS Calibur (BD). Data was analyzed using FlowJo software (Tree Star).

Flow cytometric staining of transcription factors

For intracellular T-bet staining, the Foxp3 staining buffer set (Ebioscience) was used according to the manufacturer's instructions for fixation and permeabilization. Staining was performed with an Alexa Fluor 647-coupled anti-T-bet antibody (Ebiosciene, 4B10) and with the appropriate isotype control antibody (Ebioscience). The mean relative T-bet expression level was estimated by subtracting the mean background (isotype control staining) from the mean T-bet signal. For pStat4 staining, the cells were fixed using Phosflow Lyse/Perm buffer (BD) and permeabilized with Perm Buffer III (BD) according to the manufacturer's instructions. Staining was performed with an Alexa Fluor 647-conjugated anti-pStat4 antibody (BD).

12.4 Quantitative PCR

Quantitative RT-PCR allows measuring the amount of a specific mRNA in a sample. Total RNA was isolated from $0.5 - 1 \cdot 10^6$ T-cells ($5 \cdot 10^6$ for naive cells) using Nucleospin RNA II (Macherey-Nagel) according to manufacturer's instructions. With TaqMan® Reverse Transcription Reagents (Applied Biosystem) 300-400 ng of total RNA was transcribed to cDNA in a total volume of 20 µl. Real-time PCR was performed using a LightCycler® 2.0 Instrument (Roche), in a total volume of 5 µl with LightCycler® Fast-Start DNA Master SYBR Green I. Primer sequences and reaction conditions are given in Table. 12.1. DNA was denatured for 10 min at 96°C, followed by 42 cycles of melting (15 s, 96°C), primer annealing (12 s, 60/65°C) and elongation (12 s, 72°C). The specificity of each reaction was controlled by a melting curve measured through a ramp from 60°C to 96°C with 0.1 C°/sec. The relative amount of mRNA was calculated from the crossing points (c.p.) as $E^{c.p\,input} - E^{c.p\,sample}$, where E represents the reaction efficiency, determined by serial dilution of DNA.

12.5 Chromatin immunoprecipitation

In this method, a DNA-binding protein of interest is precipitated with a specific antibody, after protein and DNA have been crosslinked in the cell. The amount of precipitated DNA is determined by qPCR. Cells were fixed with dithiobis(succinimidyl propionate) (2 mM final conc.) for 30 min at room temperature and fixation was stopped by addition of glycine (125 mM final concentration) for 10 min. Then, cells were fixed with 1 % Formaldehyde for 10 min at room temperature. The chromatin was sheared to 200–1000 bp of length by sonication with five pulses of 10 s at 30 % power (Bandelin, Berlin, Germany). Chromatin was incubated with anti-STAT4 (Santa Cruz, C-20) antibodies overnight, followed by incubation with protein A microbeads for 1 h. Washing steps were performed on l-columns (Miltenyi Biotec). Washing

Table 12.1. Primer sequences used for qPCR and ChIP assay

mRNA	forward	reverse	PCR conditions
IFN-γ	CAACAACATAAgCgTCATT	ATTCAAATAgTgCTggCAgA	60°C, 2mM $MgCl_2$
IL-12Rβ2	CTgATCCTCCATTACAgAA	CggAAgTAACgAATTgAgAA	60°C, 2mM $MgCl_2$
T-bet	TCCTgCAgTCTCTCCACAAgT	CAgCTgAgTgATCTCTgCgT	65°C, 3mM $MgCl_2$
GATA3	CCTACCgggTTCggATgTAAgT	AgTTCgCgCAggATgTCC	65°C, 3mM $MgCl_2$
STAT4	AACCCTCCATCTgTCACTTT	CCATgATgTACCCATCAATC	60°C, 3mM $MgCl_2$
Hlx	CCTTAAgCTCCAACCCAAgA	AACCTCTTCTCCAggCCTTT	62°C, 3mM $MgCl_2$
Runx3	gCCggCAATgATgAgAACTA	TCCATCCACAgTgACCTTgA	60°C, 3mM $MgCl_2$
HPRT	gCTggTgAAAAggACCTCT	CACAggACTAgAACACCTgC	65°C, 3mM $MgCl_2$
IFN-gR2	CCgAgTgAAgTACTggTTTC	gTgTTTggAgCACATCATC	60°C, 2mM $MgCl_2$
ChIP Primers			
Tbx21 enhancer	gAggTgTTTCCTCgAAgCTg	AgggACAggAAggCAggTAg	60°C, 2mM $MgCl_2$
Tbx21 control (-3kb)	CTAACAgggTgTgTgACTTCTgA	AgAAgAAgAACCCCgACAgTg	60°C, 2mM $MgCl_2$
IL-12Rβ2 (-2.2kb)	AAAgCACgTCAgAAATggAA	TAAggCAACAgAAgCCCAAC	60°C, 3mM $MgCl_2$
IL-12Rβ2 (-0.7kb)	TCAAgTgggCCTggAATTTA	ggCTCAgAgCTgTggAAATC	60°C, 3mM $MgCl_2$
IL-12Rβ2 (-0.2kb)	gTgAgTTgTTTAgCgCgTCA	AgAAAggAACCAAggCgTgT	60°C, 3mM $MgCl_2$

was performed sequentially with high-salt[1], low-salt[2], LiCl[3], and TE[4] buffer. The chromatin precipitate was eluted with 1 % SDS, 0.1 M NaHCO3. Cross-links were reversed by incubation at 65°C for 4 hrs in the presence of 0.2 M NaCl, and the DNA was purified with NucleoSpin Extract II (Macherey-Nagel). The amount of immunoprecipitated DNA was determined by realtime PCR with a LightCycler (Roche Applied Science, Mannheim, Germany) using FASTStart SYBR Green Master (Roche Applied Science). The relative amount of DNA was calculated from the crossing points (c.p.) as $E^{c.p input} - E^{c.p sample}$, where E represents the reaction efficiency, determined by serial dilution of DNA. Primer sequences are given in Table 12.1).

12.6 Immunoblot

To measure NFAT phosphorylation, western blot quantification was used. For detection, three tandem HA-repeats at the N-terminus of the construct were used. Cells were cultured overnight in 12-well plates. After appropriate stimulation, wells were washed once with PBS, then 120 µl whole cell lysis buffer[5]was added to each well. The lysis buffer contained 1 % of the detergent NP-40 to lyse the cells and different phosphatase and protease inhibitors to prevent degradation and dephosphorylation of NFAT. The plate was incubated for 30 min at 4°C on a shaker. Lysates were centrifuged (20 000xg) for 15 min at 4°C and the pellet was discarded. Protein concentration in the sample was quantified through the Bradford assay (Biorad)

On a 14x8 cm 7.5 % polyacrylamide gel, 30 µg protein were applied per lane. The gel was run at 30 mA for ~4 hrs. Proteins were tranferred using wet-blot (Biorad) over night at 4°C at 40 V onto a nitrocellulose membrane (0.45 µm). To test for homogeneous transfer the membrane was stained with Ponceau red for 1 min, which was then removed by extensive washing with water and with TBS-T (TBS+0.1 % Tween-20). The blot was blocked with 5 % milk in TBS-T for 30 min. The primary monoclonal anti-HA antibody (12CA5) that was used to detect NFAT was applied for 4 hrs at room temperature in TBS-T+5% milk. After washing (4x 5 min TBS-T), the secondary antibody was applied: anti-mouse-HRP (Sigma-Aldridge), 1:10000 in TBS-T+5 %milk). The blot was incubated at room temperature for 1 h

[1]0.1 % SDS, 1 % Triton-X100, 2 mM EDTA, 20 mM Tris-HCl, pH 8.1, 150 mM NaCl

[2]0.1 % SDS, 1 % Triton-X100, 2 mM EDTA, 20 mM Tris-HCl, pH 8.1, 500 mM NaCl

[3]0.25 mM LiCl, 1 % IGEPAL-CA630, 1 % Desoxycholic acid, 1 mM EDTA,120 mM Tris-HCl, pH 8.1

[4]1 mM EDTA, 10 mM Tris-HCl, pH 8.1

[5]50 mM Tris, pH 7.4, 150 mM NaCl, 2 mM EGTA, 2 mM EDTA, 10 % glycerol, 1 % NP-40. Added fresh: 10 mM β-Glycerolphosphate, 5 mM NaF, 1 mM Vanadate, PMSF, Leupeptin, Apoprotinin, 20 mM $NaPP_i$, 5 mM EDTA.

and washed 4 x 5min with TBS-T. HRP was visualized by chemiluminescence (Amersham Biosciences) and photographed with Fujifilm Las-3000.

For image analysis ImageJ software was used. Individual lanes were selected and lane profiles were analyzed: The area under each band was used to calculate the NFAT fractions in the phosphorylated and dephosphorylated states.

12.7 Microscopy

To quantify NFAT localization, an NFAT construct, fused to GFP on its C-terminus was used (Aramburu et al., 1998). Cells were cultured overnight in 96-well plates. After appropriate stimulation, cells were permeabilized for 10 min with 0.2 % Triton-X100 in PBS at room temperature. To stain the cytoplasm, cells were incubated with phalloidin-Texas Red (Invitrogen, 1:20) in PBS for 20 min at room temperature. To stain the nucleus, cells were incubate for 5 min with DAPI also at room temperature. Cells were washed three times with PBS and images were acquired using ImageXpress^{MICRO} (Molecular Devices). Automated image analysis was performed with the software CellProfiler (Carpenter et al., 2006).

Chapter 13

Theoretical methods

13.1 Parameter optimization

All simulations were performed with Matlab software (Mathworks). For both models described in this thesis, simulated annealing was used for global optimization of the parameter values (Kirkpatrick et al., 1983). For implementation, a Matlab-based program was used[1]. Starting from an initial guess of parameters, the parameter space around this guess is explored and the goodness of the fit is estimated for each new parameter set, for example as the weighed the sum of squared residuals (χ^2). If a new parameter yields a better fit, it is accepted as the new optimized solution. If the new parameter set results in a decrease of the goodness of the fit, it can also be accepted as the new optimized fit with a certain probability p that depends on the worsening of the fit and on the so-called "temperature" (T) with

$$p = e^{\frac{\chi^2(old\,par) - \chi^2(new\,par)}{T}} \tag{13.1}$$

T is a control parameter that is slowly decreased during the optimization process. A high temperature allows a global scanning of the parameter space, while at the reduced temperature, the solution is more likely to stay in an optimum that was found at higher temperatures. To generate a new parameter set from an old set, for one randomly chosen parameter, a new value was calculated, using the following function:

$$p_{new} = p_{old} + (p_{old} + 0.01) \cdot a_b \tag{13.2}$$

where a_b denotes a random number drawn from a normal distribution with mean 0 and standard deviation b. Two control parameters were used to adapt the algorithm to the specific optimization problem: b and the initial

[1]The program written by Joachim Vandekerckhove was downloaded from `www.mathworks.com/matlabcentral/fileexchange/10548`.

temperature T_{ini}. For global optimization b was usually set to 0.7 and T_{ini} to 100, for local optimization b was set to 0.1 and T_{ini} to 1. Simulated annealing was usually followed by another optimization step, using a non-linear least squares algorithm (`lsqnonlin` in Matlab).

13.2 Confidence intervals

For the Th1 gene network, confidence intervals were estimated through a bootstrapping approach, where new data sets were generated from the experimental data by addition of noise. From the distribution of parameter values determined by fitting each data set, the 5th and 95th percentile was calculated to estimate the 90% confidence interval. For the NFAT model, another method was used to estimate confidence intervals, described by Raue et al. (2009) and described in detail in the Appendix B.7. For all correlation analyses presented in part I, 95 % confidence intervals for correlation coefficients were estimated from a distribution of correlation coefficients of 100 data sets obtained by a bootstrapping strategy with replacement. The interval was bias corrected using the activated percentile method (Efron and Tibshirani, 1993).

13.3 Control coefficients

Control coefficients allow to estimate, how strong a perturbation δ of a system parameter affects a property A of this system. Normalized control coefficients (cc) are calculated as

$$cc = \frac{pp}{A} \cdot \frac{\delta A}{\delta p} \tag{13.3}$$

13.4 Statistical analysis

Statistical analysis was performed using Matlab (Mathworks). In Figure 3.3, a paired t-test was applied to the logarithm of the expression levels to assess the significance of the measured relative increase of T-bet and IL-12R2 in the absence of CD3-stimulation or after addition of CsA. Log-normal distribution was verified by the Lilliefors test.

Part V

Appendix

Appendix A

Th1 network: additional material

A.1 IFN-γR2 is rapidly down-regulated

It has been described previously that the IFN-γR2 chain is downregulated in a negative feedback though IFN-γ signaling (Pernis et al., 1995; Bach et al., 1995; Liu and Janeway, 1990). To understand whether this regulatory interaction had to be considered in the Th1-inducing gene-regulatory network, expression kinetics of IFN-γR2 mRNA were quantified during Th1 differentiation (Fig. A.1). It was found that already 10 hrs after onset of the stimulation, mRNA levels were drastically downregulated and remained at a constant low expression level for the following days. Since the Th1-differentiation network operates on a slower time scale than IFN-γR2 down-regulation, IFN-γR2 was not explicitly considered in the analysis.

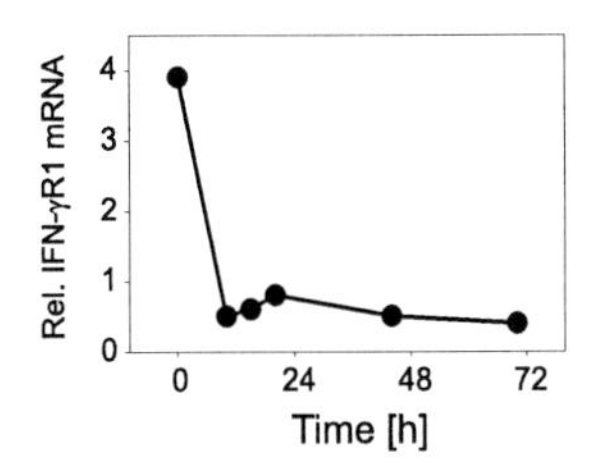

Figure A.1. IFN-γR2 expression kinetics. Naïve Th-cells, isolated from Balb/c mice were cultured under Th1-inducing conditions. IFN-γR2 mRNA kinetics were quantified.

A.2 Modeling transcriptional regulation

Transcriptional activation is controlled by binding of a transcription factor (TF) to its bindings site in the promoter of the target gene. In the presented mathematical description it is assumed that binding site occupancy in the population is proportional to the transcription rate. If the rate constants of the association and the dissociation reactions are k_{on} and k_{off}, respectively, formation of the transcription factor-DNA complex (TF-DNA) can be described as

$$\frac{\mathrm{d}[TF-DNA]}{\mathrm{d}t} = k_{on} \cdot [TF] \cdot [DNA] - k_{off} \cdot [TF-DNA] \quad \text{(A.1)}$$

This binding reaction equilibrates rapidly (~sec) compared to the process analyzed (~hours/days) and can therefore be assumed to be in the steady state. If we now take into account that the sum of the fractions of free ($[DNA]$) and occupied binding sites ($[TF-DNA]$) in the population always add up to unity, the fraction of occupied binding sites is given by

$$[TF-DNA] = \frac{[TF]}{\frac{k_{off}}{k_{on}} + [TF]} \quad \text{(A.2)}$$

In the model, all signal-dependent transcription rates were assumed to be proportional to the binding site occupancy (TF-DNA) and the signal strength was taken to be proportional to the concentration of the transcription factor ([TF]) activated downstream of that signal. Therefore, all signal-dependent transcription rates in the model were modeled as saturating, hyperbolic functions of the inducing signals, as shown in equation A.2.

A.3 No NFAT binding to the *Il12rb2* promoter

Cell isolation and stimulation was performed by Vladimir Pavlovic and the precipitation of the chromatin was performed by Hyun-Dong Chang.

It has been reported previously for human T-cells that NFAT1 binds to the *Il12Rb2* promoter 200 bp upstream of the transcriptional start site (van Rietschoten et al., 2001). Sequence analysis of the region upstream of the murine *Il12rb2* gene showed that this binding site was not conserved in mice (Fig. A.2A, red and green). Instead, NFAT binding was predicted at 0.7 and 2.2 kb upstream of the start site. In particular, the -2.2 kb binding site was located in a region with high cross-species conservation (Fig. A.2A, blue track). To test, whether NFAT1 would bind to any of the three

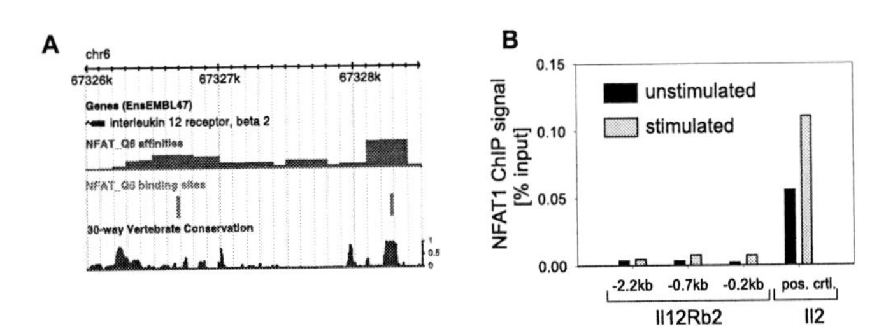

Figure A.2. NFAT1 does not bind to the ***Il12Rb2*** **promoter.** (A) Sequence analysis of a 2.5kb region upstream of the transcriptional start site of the murine *Il12Rb2* gene (black) was performed, using the promotion genome browser (http://promotion.molgen.mpg.de). Two different methods were used to predict NFAT binding sites: Conventional matrix-based binding site prediction (green) and calculation of binding affinity based on a physical model (red) (Manke et al., 2008). Both methods predicted a potential weak binding site at -0.7 kb and a stronger binding site at -2.2 kb, where only the latter was located in a region of high 30-way vertebrate sequence conservation (blue). (B) The two predicted binding sites (-2.2, -0.7) and a previously published binding site at -0.2 kb (van Rietschoten et al., 2001) were tested for NFAT1 binding by ChIP assay. (B) $CD4^+$ T-cells isolated from Balb/c mice were either left unstimulated or they were stimulated with ionomycin, which activates NFAT. While no binding to the *Il12rb2* promoter could be detected, NFAT did bind to a known binding site in the *Il2* promoter, which was increased by the stimulation.

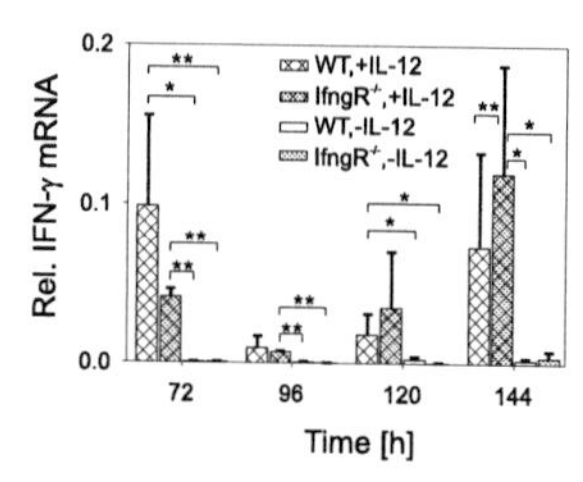

Figure A.3. Low-level IFN-γ expression in response to IL-12. (A,B) Naïve Th-cells from C57BL/6 or *Ifngr$^{-/-}$* mice were stimulated in the presence of IL-12 or IFN-γ and blocking antibodies to IL-12. IFN-γ mRNA was quantified in five independent experiments, mean and s.d. are shown. The data was statistically analyzed at time points 72-144 h. A pairwise comparison of the expression levels under different conditions was performed with a paired t-test. * and ** indicate that expression levels were significantly different with $p<0.01$ and $p<0.005$, respectively.

candidate binding sites (-0.2, -0.7, -2.2), $CD4^+$ T-cells were isolated from mice and stimulated for three hours with ionomycin that activates NFAT1. Chromatin-Immunoprecipitation was performed with a NFAT1-specific antibody and binding was readily detected at a known NFAT binding site in the *Il2* promoter, where it was further enhanced through stimulation (Fig. A.2B, Il2). By contrast, no NFAT1 binding was detected to any of the three sites in the Il12Rb2 promoter (-0.2, -0.7, -2.2) (Fig. A.2B).

A.4 Supplementary Tables and Figures for Part I

No the following pages number supplementary material can be found for part I of this thesis, analyzing the gene-regulatory network controlling Th1 differentiation

Table A.1. Parameter values for multiple experiments

Activation rate constants	experiment 1 (blue)	experiment 2 (red)	experiment 3 (green)
α_1	0.044 h^{-1}	0.027 h^{-1}	0.037 h^{-1}
α_2	0.42 h^{-1}	0.41 h^{-1}	0.55 h^{-1}
α_3	0.00051 h^{-1}	0.0015 h^{-1}	0.038 h^{-1}
α_4	0.0028 h^{-1}	0.0011 h^{-1}	0.0004 h^{-1}
α_5	3.7 h^{-1}	2.2h^{-1}	6.5h^{-1}
Half-saturation constants			
K_1	0.46	0.014	0.093
K_2	2.1	4.9	2.8
K_4	0.013	0.0012	0.0020
K_5	0.029	0.66	4.08
K_6	66	40	2.5
K_7	0.014	0.01	0.029
s.s.r.	0.24	0.30	0.06

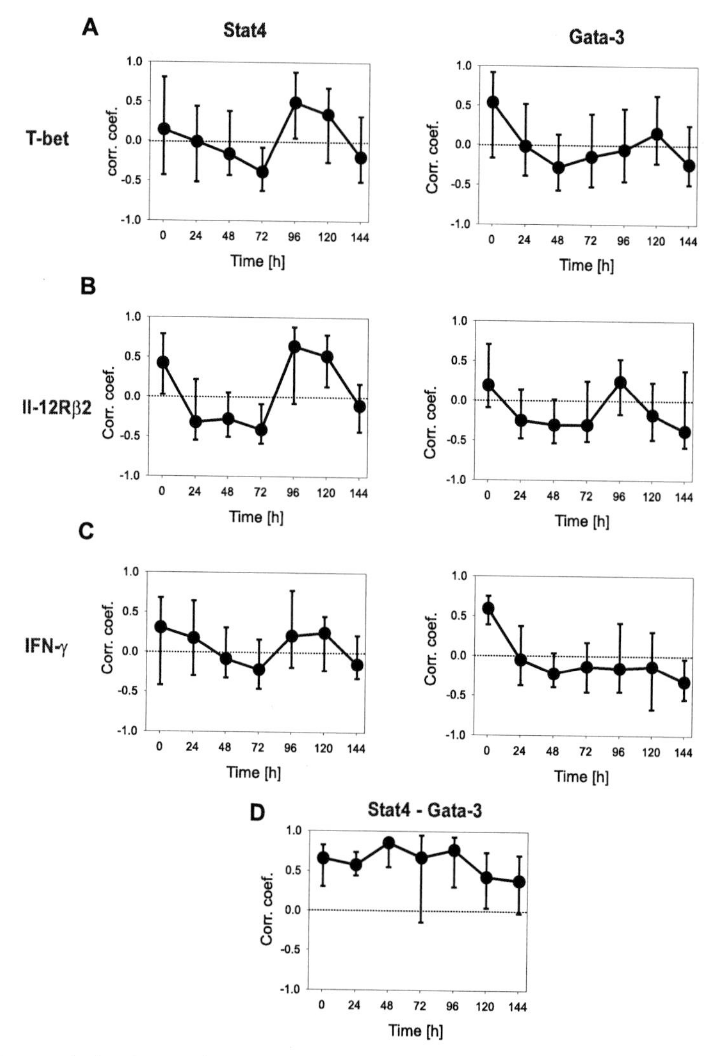

Figure A.4. STAT4 and GATA3 levels do not correlate with T-bet, IL-12Rβ2 or IFN-γ expression, but with each other. As described in Fig. 3.4, data from five experiments, each employing four different culture conditions, were used for a correlation analysis. For each time point, the correlation between STAT4 (left column) or GATA3 (right column) and T-bet (A) or IL-12Rβ2 (B) or IFN-γ (C) was analyzed. (D) Correlation between STAT4 and GATA3 levels. 95% confidence intervals of the plotted correlation coefficients were estimated by a bootstrapping strategy.

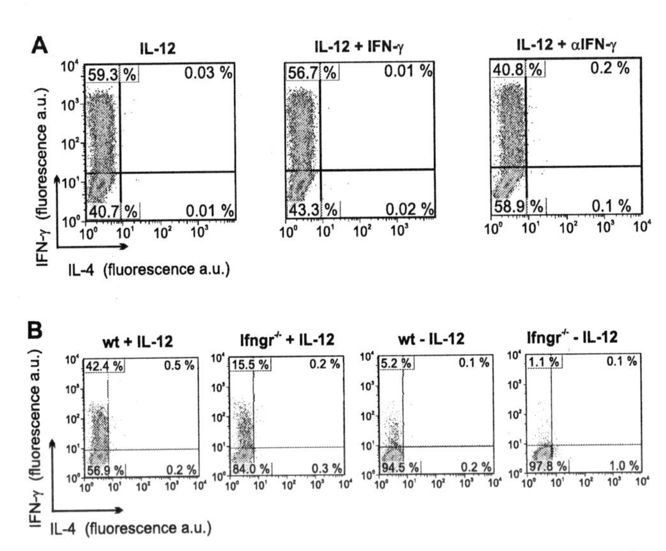

Figure A.5. IFN-γ staining. (A) Naïve Th cells ($CD4^{+}CD62L^{hi}CD44^{lo}$ purified by flow cytometry), isolated from Balb/c mice, were stimulated for six days with IL-12 and IFN-γ signaling remaining unperturbed, with IL-12 and recombinant IFN-γ or with IL-12 and blocking antibodies to IFN-γ. Cells were re-stimulated with PMA and ionomycin after six days of culture and stained intracellularly for IFN-γ and IL-4. (B) Naïve Th cells, isolated from C57BL/6 or *Ifngr*$^{-/-}$ mice, were stimulated with IL-12 or with IFN-γ and blocking antibodies to IL-12. After six days, cells were re-stimulated with PMA and ionomycin and stained intracellularly for IFN-γ and IL-4.

Table A.2. Model Equations. For the one-loop model all equations are shown. For the other models only the equations that differ from the one-loop model are depicted.

One-loop model

$$\frac{d\,Tbet_{\text{mRNA}}}{dt} = \alpha_1 + \alpha_2 \cdot \frac{Ag(t)}{K_1+Ag(t)} \cdot \frac{IFN_{\text{Prot}}}{K_2+IFN_{\text{Prot}}} - \gamma_{\text{Tbet}} \cdot Tbet_{\text{mRNA}}$$

$$\frac{d\,Rec_{\text{mRNA}}}{dt} = \alpha_4 \cdot \frac{Tbet_{\text{Prot}}}{K_8+Tbet_{\text{Prot}}} - \gamma_{\text{Rec}} \cdot Rec_{\text{mRNA}}$$

$$\frac{d\,IFN_{\text{mRNA}}}{dt} = \alpha_5 \cdot \frac{Tbet_{\text{Prot}}}{K_5+Tbet_{\text{Prot}}} \cdot \frac{Rec_{\text{Prot}}}{K_6+Rec_{\text{Prot}}} \cdot \frac{Ag(t)}{K_7+Ag(t)} - \gamma_{\text{IFN}} \cdot IFN_{\text{mRNA}}$$

$$\frac{d\,Tbet_{\text{Prot}}}{dt} = \beta \cdot Tbet_{\text{mRNA}} - \delta_{\text{Tbet}} \cdot Tbet_{\text{Prot}}$$

$$\frac{d\,Rec_{\text{Prot}}}{dt} = \beta \cdot Rec_{\text{mRNA}} - \delta_{\text{Rec}} \cdot Rec_{\text{Prot}}$$

$$\frac{d\,IFN_{\text{Prot}}}{dt} = \beta \cdot IFN_{\text{mRNA}} - \delta_{\text{IFN}} \cdot IFN_{\text{Prot}}$$

Two-loop model

$$\frac{d\,Tbet_{\text{mRNA}}}{dt} = \alpha_1 + \alpha_2 \cdot \frac{Ag(t)}{K_1+Ag(t)} \cdot \frac{IFN_{\text{Prot}}}{K_2+IFN_{\text{Prot}}} + \alpha_3 \cdot \frac{Rec_{\text{Prot}}}{K_3+Rec_{\text{Prot}}} - \gamma_{\text{Tbet}} \cdot Tbet_{\text{mRNA}}$$

$$\frac{d\,Rec_{\text{mRNA}}}{dt} = \alpha_4 \cdot \frac{Tbet_{\text{Prot}}}{K_8+Tbet_{\text{Prot}}} \cdot \frac{K_4}{K_4+Ag(t)} - \gamma_{\text{Rec}} \cdot Rec_{\text{mRNA}}$$

Simplified two-loop model

$$\frac{d\,Tbet_{\text{mRNA}}}{dt} = \alpha_1 + \alpha_2 \cdot \frac{Ag(t)}{K_1+Ag(t)} \cdot \frac{IFN_{\text{Prot}}}{K_2+IFN_{\text{Prot}}} + \alpha_3 \cdot Rec_{\text{Prot}} - \gamma_{\text{Tbet}} \cdot Tbet_{\text{mRNA}}$$

$$\frac{d\,Rec_{\text{mRNA}}}{dt} = \alpha_4 \cdot Tbet_{\text{Prot}} \cdot \frac{K_4}{K_4+Ag(t)} - \gamma_{\text{Rec}} \cdot Rec_{\text{mRNA}}$$

Extended Two-loop model

$$\frac{d\,Tbet_{\text{mRNA}}}{dt} = \alpha_1 + \frac{IFN_{\text{Prot}}}{K_2+IFN_{\text{Prot}}} \cdot \left(\alpha_2' + \alpha_2 \cdot \frac{Ag(t)}{K_1+Ag(t)}\right) + \alpha_3 \cdot \frac{Rec_{\text{Prot}}}{K_3+Rec_{\text{Prot}}} - \gamma_{\text{Tbet}} \cdot Tbet_{\text{mRNA}}$$

$$\frac{d\,IFN_{\text{mRNA}}}{dt} = \alpha_5 \cdot \frac{Rec_{\text{Prot}}}{K_6+Rec_{\text{Prot}}} \cdot \left(\alpha_5' + \frac{Tbet_{\text{Prot}}}{K_5+Tbet_{\text{Prot}}} \cdot \frac{Ag(t)}{K_7+Ag(t)}\right) - \gamma_{\text{IFN}} \cdot IFN_{\text{mRNA}}$$

Appendix B

NFAT model: Supplementary Material

B.1 Quantification of NFAT phosphorylation

To test, if the fraction of de-phosphorylated NFAT could be quantified with reasonable accuracy, two samples, containing phosphorylated and partially dephosphorylated NFAT were mixed at different ratios and the dephosphorylated fraction was quantified by western blot (Fig. B.1A-C). Since a linear relationship was observed between the mixing ratio and the quantification result, it was concluded that this method was well suited to reliably quantify the dephosphorylated fraction in a sample (Fig. B.1D).

B.2 Quantification of NFAT localization

To quantitatively assess the amount of NFAT in the nucleus and in the cytoplasm, fluorescence microscopy was used with a NFAT construct, fused to GFP. Images were acquired in a 96-well format using ImageXpress^{MICRO} (Molecular Devices). Automated image analysis was performed with the software cell profiler (Carpenter et al., 2006): In a first step, the image was compressed by subtracting the background signal (compare Fig. B.2A+B). Then DAPI staining was used to identify the nuclei. The cytoplasm was identified through phalloidin-Texas-red staining (Fig. B.2C). Typically, in stimulated cells ~60 % of the GFP fluorescence was nuclear, while in resting cells the signal was reduced to ~30 %.

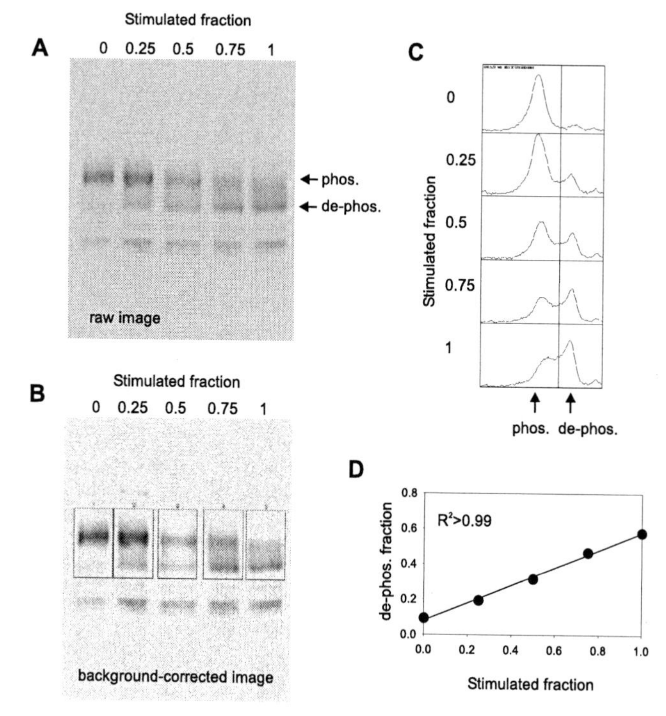

Figure B.1. Quantification of NFAT's phosphorylation status. (A) Hela-NFAT(1-460)-GFP cells were left unstimulated or cultured for 20 minutes with 6μM Ionomycin. Lysates, prepared from stimulated and resting cells, were mixed with different ratios, as indicated. NFAT phosphorylation status in these samples was analyzed by western blot with a HA-specific antibody. In the stimulated sample, the dephosphorylated fraction of NFAT is shifted towards lower molecular weight. (B) For quantification the image was background corrected, using the ImageJ software. (C) Signal intensity was integrated horizontally over each lane and plotted. A border between the phosphorylated and the de-phosphorylated fraction was defined (vertical line) and the area under the curve was used to quantify the signals. (D) A linear relationship between the fraction of stimulated lysate in the sample and the detected dephosphorylated fraction was confirmed.

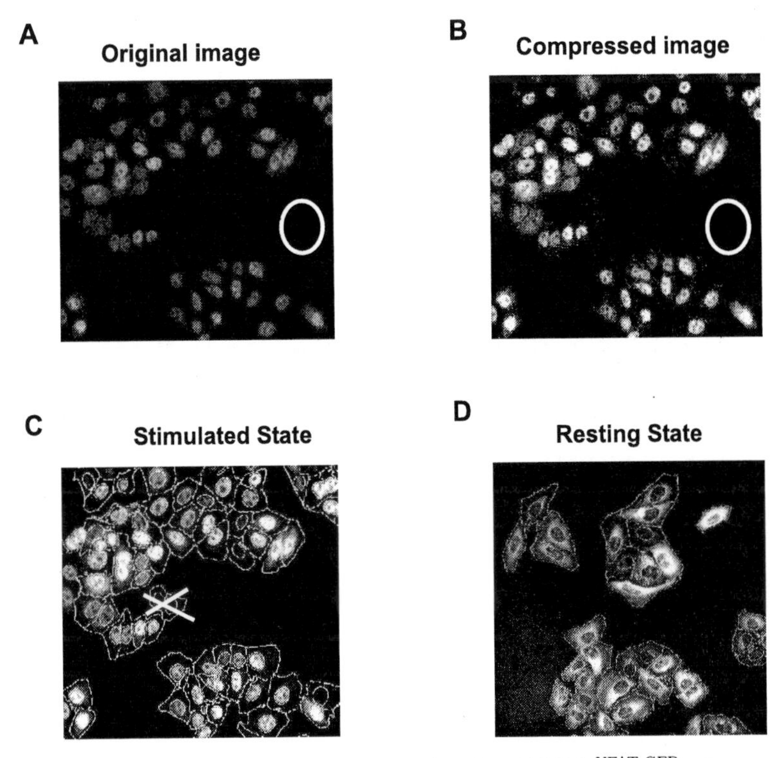

Figure B.2. Quantification of nuclear translocation. (A) Hela$^{\text{NFAT-GFP}}$ cells were stimulated for 20 min with 2 µM ionomycin and GFP fluorescence was detected by fluorescence microscopy. (B) The image in (A) was compressed through subtracting the background signals. (C) Nucleus and cytoplasm was identified and cells with very high or very low (10 %) GFP signal were eliminated from the analysis. (D) The GFP signal in unstimulated cells.

B.3 Modeling nuclear transport

In the NFAT model, it is assumed that NFAT exists in a fast conformational equilibrium between an importable and an exportable conformation, described by the equilibrium constant

$$K = \frac{[\text{Import conformation}]}{[\text{Export conformation}]} \tag{B.1}$$

The equilibrium constants differ between the phosphorylation states. It is further assumed that the import conformation can only be imported with the rate constant γ_0, but not be exported, and that the export conformation can only be exported with rate constant ε_0 and not be imported. If we want to describe the transport reactions of a nuclear species y and a cytoplasmic species x, x is described by the following equation

$$\frac{\mathrm{d}x}{\mathrm{d}t} = -\gamma_0 \cdot [\text{Imp.conf.}]_x + \varepsilon_0 \cdot [\text{Exp.conf.}]_y \tag{B.2}$$

If we take into account Equation B.1 and consider that

$$[\text{Imp.conf.}]_x + [\text{Exp.conf.}]_x = x \quad \text{and} \quad [\text{Imp.conf.}]_y + [\text{Exp.conf.}]_y = y \tag{B.3}$$

the transport reaction is described by

$$\frac{\mathrm{d}x}{\mathrm{d}t} = -\gamma_0 \cdot \frac{K \cdot x}{1+K} + \varepsilon_0 \cdot \frac{y}{1+K} \tag{B.4}$$

Through an analogous calculation the equation describing nuclear export of the nuclear species y can be found to be

$$\frac{\mathrm{d}y}{\mathrm{d}t} = \gamma_0 \cdot \frac{K \cdot x}{1+K} - \varepsilon_0 \cdot \frac{y}{1+K} \tag{B.5}$$

This equation holds only true if the nuclear and the cytoplasmic compartments are of identical sizes. However, in Hela cells the nucleus is smaller than the cytoplasm. Therefore the increase in the nuclear concentration is stronger than the decrease in the cytoplasmic concentration, when a certain number of molecules is imported. Therefore, a correction factor

$$\rho = \frac{V_{cyt}}{V_{nuc}} \tag{B.6}$$

where V denotes the volumes of the respective compartments must be introduced in Equation B.5 describing the nuclear species y:

$$\frac{\mathrm{d}y}{\mathrm{d}t} = \rho \cdot \gamma_0 \cdot \frac{K \cdot x}{1+K} - \rho \cdot \varepsilon_0 \cdot \frac{y}{1+K} \tag{B.7}$$

In this way, the transport reaction of NFAT in Table 10.1 were derived.

Table B.1. Parameter constraints

Parameter	Upper bound	Lower bound
α, β [min^{-1}]	0.01	10
δ_{cyt} [min^{-1}]	0.2	10
δ_{nuc} [min^{-1}]	0	1
δ_0 [%]	0	100
K	0	1
γ_0 [min^{-1}]	0.01	10
ε_0 [min^{-1}]	0.01	10
ω [%]	0	100
κ	0	10
λ	1	10

B.4 The ionomycin dose-response curve

To simulate the ionomycin dose-response curve it must be known, how a change in the ionomycin concentration affects calcineurin activity. However, it is not possible to measure this relationship in cells. Therefore, a generic function was used to describe this relations ship and its parameter values were estimated from fitting the model to the data. The following function was used to describe, how the phosphatase activity δ^* depended on the ionomycin concentration

$$\delta^*([Iono]) = \frac{\kappa \cdot [Iono]^\lambda}{\nu + [Iono]^\lambda} \tag{B.8}$$

A concentration of 2 µM was as a reference, where the calcineurin activity was given by δ. With

$$\frac{\kappa \cdot [2\mu M]^\lambda}{\nu + [2\mu M]^\lambda} \equiv 1 \tag{B.9}$$

one parameter ν could be eliminated and equation B.8 was reduced to

$$\delta^* = \delta \cdot \frac{\kappa \cdot (0.5 \cdot [Iono])^\lambda}{\kappa - 1 + (0.5 \cdot [Iono])^\lambda}$$

B.5 Model Fitting

To fit the NFAT model to the experimental data, a multi-step optimization procedure was applied. In the first step, a simulated annealing algorithm was used, described in Section 10.2, to globally search the parameter space, starting from an initial temperature T_{ini}=100. The standard deviation b of the normally distributed random numbers that were used to generate new parameter sets was set to b=0.7. The weighed sum of squared residuals (χ^2) was used as an estimator of the goodness of the fit. The parameter constraints are shown in Table B.1. The full model with 27 parameters was fitted 120 times, starting from random parameter values generated as

$$par = e^r \tag{B.10}$$

where r denotes a random number drawn from a normal distribution with mean 0 and s.d. 1. Phosphorylation and localization of NFAT during the kinetic and dose response experiments with different kinases inhibitors were simulated simultaneously and fitted to the experimental data. The best ten parameter sets were then further optimized with the simulated annealing algorithm: For a local optimization, $T_{init}$was set to 1 and b was set to 0.1 and the parameter set was optimized 20 times, always starting from the previous best fit. The best five parameter sets are shown in Table B.2, each derived independently from random starting values. The best fit of the full model is shown in Fig. B.3.

In the model selection procedure, described in the next section, a similar optimization procedure was used. Global optimization was performed 100 times from random starting values, then the best fit was further optimized locally. In parallel, the parameter set found for the full model was used as an initial parameter set and was then further optimized for the specific model investigated.

B.6 Model Selection

In the used step-up model selection procedure one starts with a simple model and extends the model further until no significant improvement can be detected. To compare models of distinct complexities, a likelihood ratio test was applied (Timmer et al., 2004) using the weighed sum of squared residuals as an estimator of the goodness of the fit (χ^2). The fact that this estimator can be assumed to follow a χ^2 distribution, was used for statistical model comparison. The log-likelihood ratio was calculated as the difference in χ^2 between the two compared model. To test if the more complex model provided a significant improvement, the log-likelihood ratio was compared to the χ^2 distribution, corresponding to the adequate number of degrees of freedom, given by the number of additional parameters in the more complex model. An decrease in χ^2 was considered a true improvement when this comparison yielded a p-value smaller than 0.1.

Table B.2. Parameter values for full model

Parameter	Fit #1	Fit #2	Fit #3	Fit #4	Fit #5
$\alpha_{cyt,1}$ [min-1]	0.022	0.028	0.027	0.026	0.015
$\alpha_{nuc,1}$ [min-1]	0.01	0.01	0.01	0.011	0.011
$\alpha_{cyt,2}$ [min-1]	9.8	10	9.8	9.9	9.3
$\alpha_{nuc,2}$ [min-1]	6.7	7.2	10	5.2	4.6
$\beta_{cyt,1}$ [min-1]	0.12	0.12	0.12	0.11	0.04
$\beta_{nuc,1}$ [min-1]	0.025	0.025	0.026	0.027	0.045
$\beta_{cyt,2}$ [min-1]	9.1	8.5	5.6	9.6	8
$\beta_{nuc,2}$ [min-1]	8.6	6.5	6.3	4.5	9.2
$\delta^{CK}_{cyt,1}$ [min-1]	0.42	0.47	0.38	0.47	0.35
$\delta^{CK}_{nuc,1}$ [min-1]	0.0041	9.3e-005	0.83	4.4e-006	0.0057
$\delta^{GS}_{cyt,1}$ [min-1]	0.77	0.83	0.62	0.94	0.7
$\delta^{GS}_{nuc,1}$ [min-1]	0.19	0.1	0.12	0.1	0.23
$\delta^{GS}_{cyt,2}$ [min-1]	0.29	0.3	0.31	0.32	0.23
$\delta^{GS}_{nuc,2}$ [min-1]	0.012	0.0099	0.0099	0.0085	0.023
$\delta^{CK}_{cyt,2}$ [min-1]	0.22	0.2	0.2	0.2	0.2
$\delta^{CK}_{nuc,2}$ [min-1]	0.54	0.69	0.49	0.38	0.36
δ_0 [%]	5.0	4.7	4.5	4.4	5.4
K_{PP}	0.03	0.034	0.042	0.04	0.045
K_{PO}	0.91	0.97	0.94	0.96	1
K_{OP}	0.02	0.022	0.027	0.026	0.025
K_{OO}	0.37	0.42	0.51	0.47	0.47
γ_0 [min-1]	0.42	0.37	0.32	0.33	0.36
ε_0 [min-1]	0.011	0.011	0.011	0.011	0.013
ω_{CK} [%]	$2.2{\cdot}10^{-4}$	$1{\cdot}10^{-4}$	$3.3{\cdot}10^{-3}$	$2.2{\cdot}10^{-3}$	$1.6{\cdot}10^{-4}$
ω_{GS} [%]	59	58	57	57	57
κ	1.3	1.3	1.3	1.3	1.3
λ	2.1	2.1	2.2	2.2	2.1
χ^2	34.5	34.6	34.6	34.8	35.1

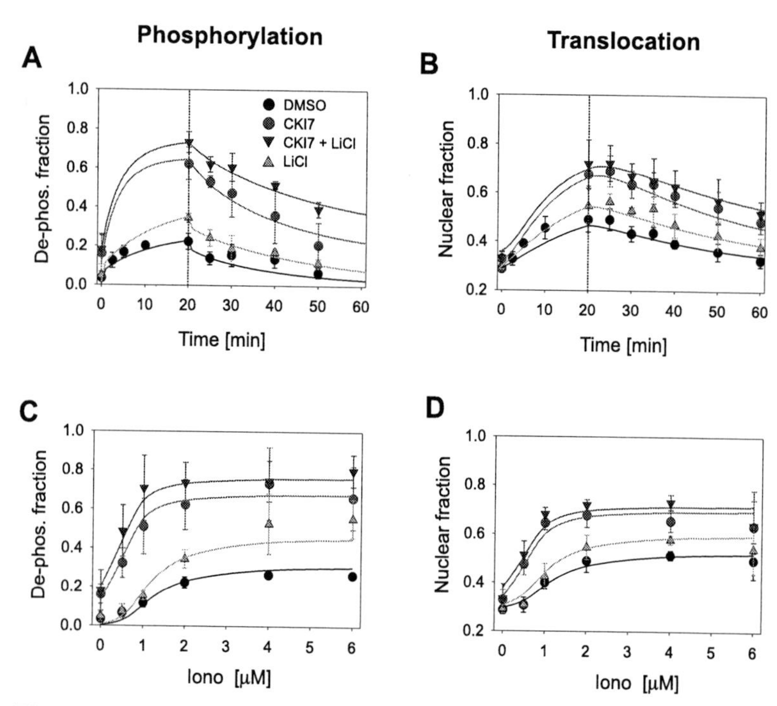

Figure B.3. Best fit of the full model. The 27 model parameters were fitted to the experimental data described in Fig. 9.2. Lines represent the model simulation and symbols the experimental data.

Table B.3. Multistep model selection procedure. The selected features are highlighted in gray

Model features	χ^2	$\Delta\chi^2$	# par.	p
	1. Model			
-	502	-	10	-
$\alpha \neq \beta$	478	24	11	<0.005
$\alpha_1 = \beta_1 \neq \alpha_2 = \beta_2$	213	289	11	<0.005
δ_{cyt} independent	441	61	13	<0.005
$\delta_{nuc} > 0$ and independent	486	16	14	<0.005
$\omega_{CK} \neq \omega_{GS}$	453	49	11	<0.005
$K_{OP} \neq K_{PO} \neq K_{PP}$	481	21	12	<0.005
	2. Model: $\alpha_1 = \beta_1 \neq \alpha_2 = \beta_2$			
$\alpha \neq \beta$	68.4	125	12	<0.005
δ_{cyt} independent	81.2	132	14	<0.005
$\delta_{nuc} > 0$ and independent	88.0	125	15	<0.005
$\omega_{CK} \neq \omega_{GS}$	53.6	159	12	<0.005
$K_{OP} \neq K_{PO} \neq K_{PP}$	124	89	13	<0.005
	3. Step new basic model: $\omega_{CK} \neq \omega_{GS}$			
$\alpha \neq \beta$	53.6	0	13	<0.9
δ_{cyt} independent	53.4	0.2	15	<0.9
$\delta_{nuc} > 0$ and independent	52.9	0.7	16	<0.9
$K_{PP} = K_{0P} \neq K_{P0} \neq K_{00}$	49.3	4.3	13	<0.05
$K_{00} = K_{P0} \neq K_{0P} \neq K_{PP}$	47.5	6.1	13	<0.025
$K_{0P} = K_{P0} \neq K_{PP} \neq K_{00}$	51.7	1.9	13	<0.25
$K_{00} = K_{0P} \neq K_{P0} \neq K_{PP}$	78	-	13	-
$K_{PP} = K_{P0} \neq K_{0P} \neq K_{00}$	50.5	3.1	13	<0.1
	4. Model: $K_{00} = K_{P0} \neq K_{0P} \neq K_{PP}$			
$\alpha \neq \beta$	41.5	6.0	15	<0.05
δ_{cyt} independent	44.6	2.9	16	<0.5
$\delta_{nuc} > 0$	46.7	0.8	14	<0.25
$K_{00} \neq K_{P0}$	47.6	0	14	<0.9
	5. Model: $\alpha \neq \beta$			
δ_{cyt} independent	40.6	0.9	18	<0.9
$\delta_{nuc} > 0$	37.7	3.8	16	<0.1
$K_{00} \neq K_{P0}$	41.1	0.4	16	<0.75
	6. Model: $\delta_{nuc} > 0$			
$\delta_{CK} \neq \delta_{GS}$	37.1	1.7	18	<0.5
$\delta_1 \neq \delta_2$	37.8	1.0	18	<0.75
$K_{00} \neq K_{P0}$	38.4	0.4	17	<0.75
all parameter independent	34.5		27	

The initial simple model (1. model) contained 10 independent parameters and made the following assumptions:

1. De-phosphorylation rates in the cytoplasm (δ_{cyt}) are identical.
2. No dephosphorylation occurs in the nucleus ($\delta_{nuc} = 0$).
3. Kinase activities are identical for all motifs and substrates ($\alpha = \beta$).
4. LiCl and CKI7 work with the same efficiency ($\omega_{CK} = \omega_{GS}$).
5. The equilibrium constants for three phosphorylated states are identical ($K_{PP} = K_{PO} = K_{OP}$).

This simple model, however, was unable to account for the experimental data ($\chi^2 = 502$, Table B.3, 1. model). In the first step of the model selection procedure, several new features were tested and the biggest improvement of the fit was reached ($\chi^2 = 213$, Table B.3, 2. model), when assuming that the first and second phosphorylation steps proceeded with different rates ($\alpha_1 = \beta_1 \neq \alpha_2 = \beta_2$). In the next step, the assumption that the two kinase inhibitors work with different efficiencies ($\omega_{CK} \neq \omega_{GS}$) resulted in a further improvement of the fit ($\chi^2 = 53.6$, Table B.3, 3. model). In a similar manner three more rounds of model selection were performed until no further improvement was possible (Table B.3). The final model ($\chi^2 = 37.7$) made the following assumptions:

1. Phosphorylation rates differ between the two motifs and depend on the state of not targeted motif, but do not differ between nucleus and cytoplasm.
2. De-phosphorylation rates are identical for the SRR-1 and SP2 motifs and are independent of the not-targeted motif.
3. De-phosphorylation activity exists in the nucleus.
4. LiCl and CKI7 exhibit different inhibition efficiencies.
5. The equilibrium constants K differ between the phosphorylation states, except that K is identical for the de-phosphorylated and the partially phosphorylated PO states.

The final model was used to simulate the kinetics of all species during activation and deactivation (Fig. B.4).

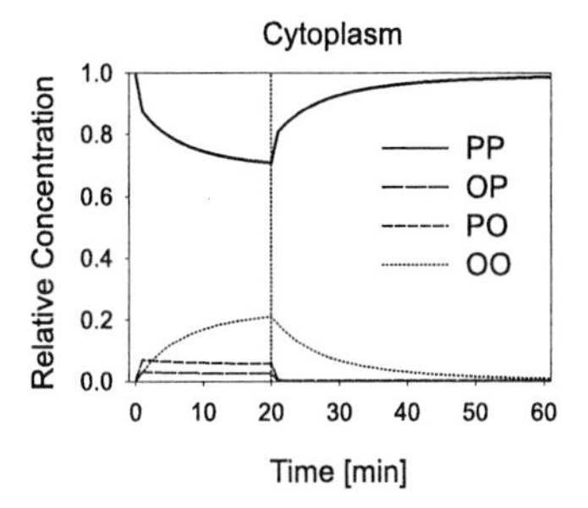

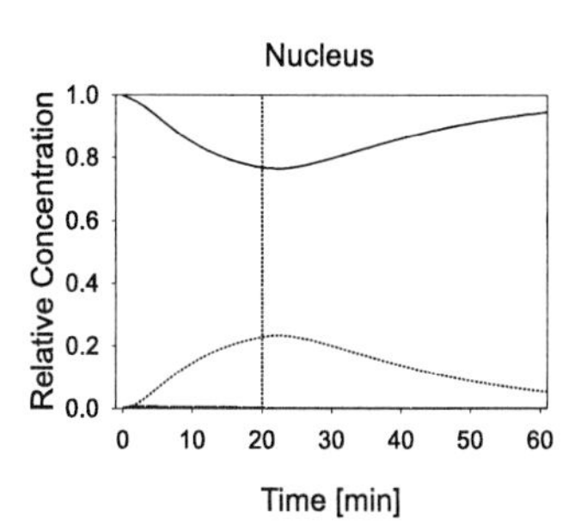

Figure B.4. Simulation of the NFAT phosphorylation states. The best fit of the final model was used to simulate the relative abundance of each phosphorylation state in the cytoplasm (left) and in the nucleus (right).

B.7 Parameter Identifiability

To understand, how reliably a parameter can be estimated from the data, it must be tested, how much the parameter value can be changed until the goodness of the fit decreases significantly. Therefore each parameter was shifted to lower and higher values in a step-wise manner, starting from the estimated value (Raue et al., 2009). For each new value, the remaining 15 parameters were re-optimized and the optimal χ^2 was calculated. In this way, a likelihood-profile for all parameters was calculated (Fig. B.5). 90 % confidence intervals were estimated using a threshold for the likelihood, derived from the χ^2 distribution (40.4) (Table 10.2). A parameter is non-identifiable, if χ^2 does not exceed this threshold also for big changes of the parameter value. None of the model parameters were completely non-identifiable, but for $\alpha_1, \delta_{nuc}, K_{PP}, K_{OP}$ and ω_{CK} only an upper bound of the parameter values could be determined, for α_2, γ and K_{XO}, which describes K_{OO} and K_{PO}, only a lower bound was well defined (Fig. B.5). But the order of magnitude could be estimated for all parameters, except for the parameters that describe nuclear import, in particular K_{XO} and γ yielded a good fit of the data over large regions of their parameter space. Moreover, the confidence interval found for K_{XO} overlaps strongly with the confidence intervals for the other equilibrium constants.

In principle, non-identifiability can arise, when parameters are correlated. Therefore, the parameter sets that yielded equally good fits were analyzed. All parameter values for K_{XO} that allowed a fit with $\chi^2 < 40.4$ were chosen, and the correlation with the re-optimized values of all other parameters were analyzed. A strong linear correlation was found with K_{OP} and K_{PP} that were

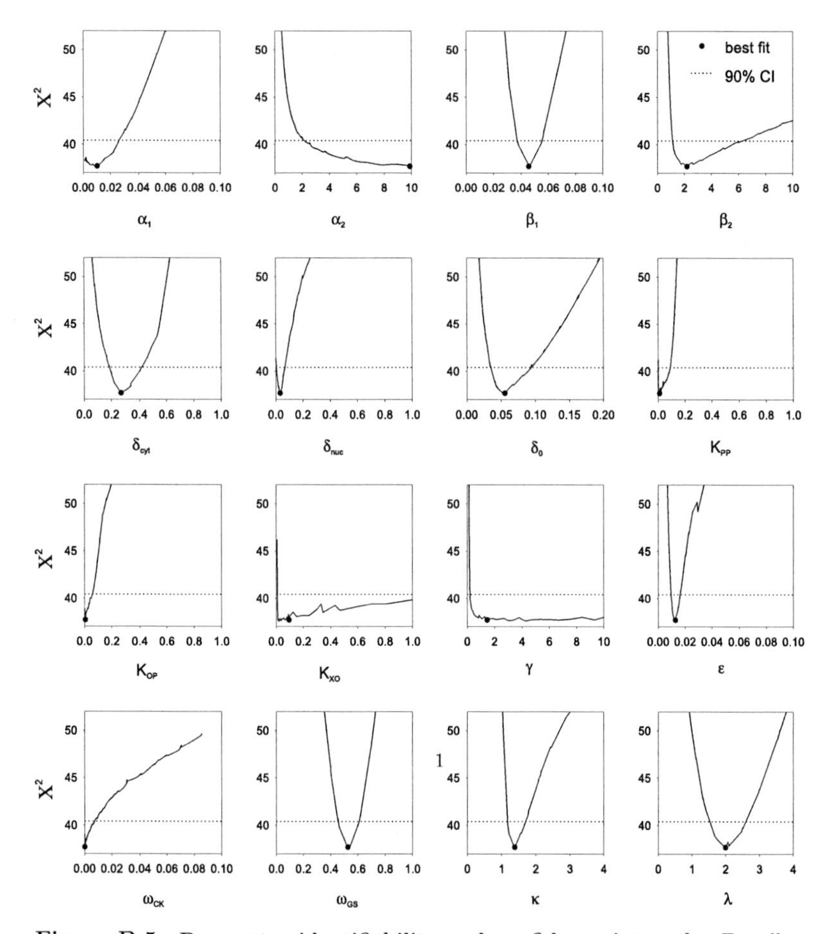

Figure B.5. Parameter identifiability and confidence intervals. For all free parameters i of the NFAT model, the profile likelihood was calculated by scanning the parameter space of i and re-optimizing χ^2. χ^2 was plotted against the value of i, the dot denotes the optimal parameter value. 90 % confidence intervals were derived using a threshold in the likelihood. The threshold was set to 2.7 above optimal χ^2 (dotted line).

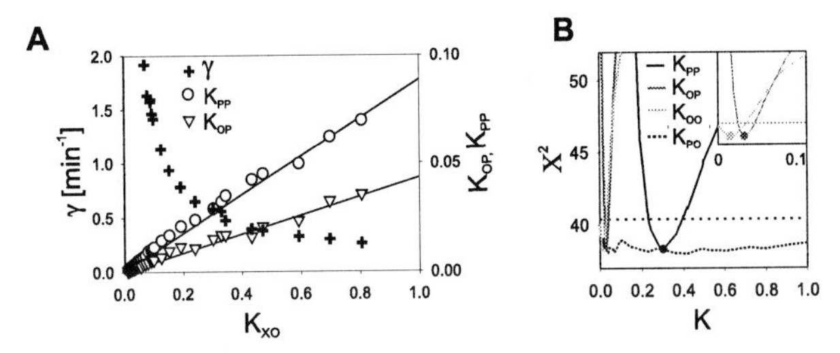

Figure B.6. Identifiability of transport parameters. (A) From the likelihood profile for K_{XO} in Fig. B.5, the parameter values were chosen, where $\chi^2 < 40.4$. The values of K_{XO} are plotted against the optimized values of K_{PP}(circles), K_{OP} (triangles) and γ_0 (crosses). For $K_{PP}$and K_{OP} a linear regression was performed and it was found that $K_{PP} \sim 0.08 \cdot K_{XO}$ and $K_{OP} \sim 0.04 \cdot K_{XO}$. (B) Likelihood profiles for the equilibrium constants, when γ_0 is fixed to 0.5 min^{-1}.

also linearly correlated with each other (Fig. B.6 circles and triangles) and an inverse hyperbolic relationship with the import rate constant γ was found (Fig. B.6 crosses). Therefore a decrease in K_{XO} seems to be compensated by an increase in the import rate γ, which then requires a decrease of the other equilibrium constants. In this way, the effective import rates of the different states stay constant, although the equilibrium constants are changed. Nevertheless, from the correlation, it can be concluded that K_{XO} must be assumed to be 10-fold higher than K_{PP} and K_{OP} to explain the data (Fig. B.6). To constrain the parameter space, a realistic, rather low value was assumed for γ, such as 0.5 min^{-1}. For the fixed value of γ, confidence intervals were again calculated for the equilibrium constants and all three were found to be well defined (Fig. B.6, right and Table 10.3). K_{PP} and K_{OP} ($\sim$0.01) could not be distinguished, but both were significantly smaller than K_{XO} ($\sim$0.3). In addition, the likelihood profile for K_{PO} was calculated, independently from K_{OO} and it was found to be non-identifiable.

B.8 Modeling DSCR1 and DYRK overexpression in Down Syndrome

Interestingly, some symptoms of trisomy 21, which causes Down syndrome, have been attributed to deregulated NFAT signaling. DSCR1, a negative regulator of calcineurin, as well as DYRK1A, another NFAT kinase that probably targets the SP3 motif, are located on the chromosome 21 and are overexpressed in down-syndrome fetuses (Arron et al., 2006; Baek et al., 2009). To understand the effect of a combined over-expression of GSK-3/DYRK1A

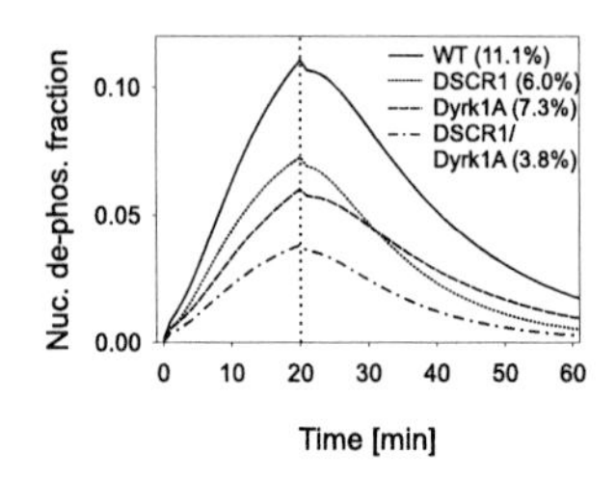

Figure B.7. Modeling DSCR1 DYRK overexpression. A 1.5-fold over-expression of DSCR1 and DYRK1A, as occurring in down syndrome cells, was simulated with the mathematical model. The kinetics of the transcriptionally active form of NFAT (de-phosphorylated in the nucleus) were simulated in response to a 20 minute Ionomycin stimulation (solid line). DSCR1 over-expression (dotted line) simulated by a 1.5-fold reduction of calcineurin activity. DYRK1A over-expression was simulated by assuming a 1.5-fold increase of GSK-3 activity, which also partially represents DYRK in the model (dashed line). In the dash-dotted line, both effects were combined.

and DSCR1, a 1.5-fold gene dosage was simulated with the mathematical model by decreasing the rate constants for the respective reactions (δ, β) 1.5-fold (Fig. B.7). In wildtype cells (solid line), the transcriptionally active fraction reached 11.1 % after 20 min of stimulation. Over-expression of GSK-3 (7.3 %) had a similar effect as reduction of calcineurin activity (6 %), resulting in a 1.85 and 1.5-fold reduction of fraction of dephosphorylated NFAT in the nucleus, respectively. A combination of both perturbations resulted in dephosphorylated nuclear fraction of 3.8 %, which is only slightly less than addition of the single effects (6/1.5=4, or 7.3/1.85=3.9). Therefore, in the model over-expression of DSCR1 and DYRK1A do not result in synergistic effects. Nevertheless, the combination results in a reduction to one third of the active NFAT levels in wildtype cells.

Table B.4. Control coefficients

Parameters		Resting state	Stimulated state
α_1	phosphorylation of the OO-state by CK1	-0.14	-0.09
α_2	phosphorylation of the OP-state by CK1	-0.12	-0.14
β_1	phosphorylation of the OO-state by Gsk3	-0.65	-0.43
β_2	phosphorylation of the PO-state by Gsk3	-1.0	-0.50
δ_{cyt}	cytoplasmic dephosphorylation at 1μM Iono	1.44	1.30
δ_{nuc}	nuclear dephosphorylation at 1μM Iono	0.19	0.08
δ_0	basal cytoplasmic dephosphorylation	1.62	0
K_{PP}	equilibrium constant between the import and the export conformation of the PP state	-0.11	-0.21
K_{OP}	equilibrium constant of the OP state	0	0
K_{XO}	equilibrium constant of the OO state	0.38	0.38
γ	nucelar import	0.30	0.19
ε	nuclear export	-0.02	0.12
α_1, α_2	CK1 activity	-0.26	-0.22
β_1, β_2	Gsk3 activity	-1.64	-0.92
$\delta_{cyt}, \delta_{nuc}$	Calcineurin activity	1.63	1.34

Abbreviations

a.u.	arbitrary units
APC	antigen presenting cell
cc	correlation coefficient
CK1	casein kinase 1
DAPI	4',6'-Diamidino-2-phenylindol, DNA stain
DYRK	dual-specificity tyrosine-phosphorylation regulated kinase
Ets1	v-ets erythroblastosis virus E26 oncogene homolog 1
Foxp3	forkhead box 3
GATA3	GATA binding protein 3
GFP	Green fluorescent protein
GFP	green fluorescent protein
GSK3	glycogen synthase kinase 3
Hlx	H2.0-like homeobox
HPRT	Hypoxanthin-Guanin-Phosphoribosyltransferase
hr(s)	hour(s)
IFN-γ	interferon-γ, cytokine
IgG2a	immunoglobulin G2a
IL-12	interleukin-12, cytokine
IL-13	interleukin-13, cytokine
IL-17	interleukin-17, cytokine

IL-4 interleukin-4, cytokine

IL-5 interleukin-5, cytokine

IL-6 interleukin-6, cytokine

JAK Janus kinase

NES nuclear export signal

NFAT nuclear factor of activated T-cells

NK cell natural killer cell

NLS nuclear localization signal

PBS phosphate-buffered saline

pSTAT4 phosphorylated STAT4 protein

RORγt RAR-related orphan receptor γt

Runx3 runt-related transcription factor 3

s.d. standard deviation

s.d. standard deviation

SRR1 serine rich region 1

Stat signal transducer and activator of transcription

T-bet T-box expressed in T-cells

TCR T-cell receptor

TGF-β transforming growth factor β, cytokine

Bibliography

Abbott, K. L., Friday, B. B., Thaloor, D., Murphy, T. J., Pavlath, G. K. 1998. Activation and cellular localization of the cyclosporine A-sensitive transcription factor NF-AT in skeletal muscle cells. *Mol Biol Cell* 9 (10), 2905–2916.

Afkarian, M., Sedy, J. R., Yang, J., Jacobson, N. G., Cereb, N., Yang, S. Y., Murphy, T. L., Murphy, K. M. 2002. T-bet is a STAT1-induced regulator of IL-12R expression in naive CD4+ T cells. *Nat Immunol* 3 (6), 549–57.

Ahn, H. J., Tomura, M., Yu, W. G., Iwasaki, M., Park, W. R., Hamaoka, T., Fujiwara, H. 1998. Requirement for distinct Janus kinases and STAT proteins in T cell proliferation versus IFN-gamma production following IL-12 stimulation. *J Immunol* 161 (11), 5893–5900.

Aramburu, J., Garcia-Cózar, F., Raghavan, A., Okamura, H., Rao, A., Hogan, P. G. 1998. Selective inhibition of NFAT activation by a peptide spanning the calcineurin targeting site of NFAT. *Mol Cell* 1 (5), 627–637.

Arron, J. R., Winslow, M. M., Polleri, A., Chang, C.-P., Wu, H., Gao, X., Neilson, J. R., Chen, L., Heit, J. J., Kim, S. K., Yamasaki, N., Miyakawa, T., Francke, U., Graef, I. A., Crabtree, G. R. 2006. NFAT dysregulation by increased dosage of DSCR1 and DYRK1A on chromosome 21. *Nature* 441 (7093), 595–600.

Assenmacher, M., Lohning, M., Scheffold, A., Richter, A., Miltenyi, S., Schmitz, J., Radbruch, A. 1998. Commitment of individual Th1-like lymphocytes to expression of IFN-gamma versus IL-4 and IL-10: selective induction of IL-10 by sequential stimulation of naive Th cells with IL-12 and IL-4. *J Immunol* 161 (6), 2825–32.

Avni, O., Lee, D., Macian, F., Szabo, S. J., Glimcher, L. H., Rao, A. 2002. T(H) cell differentiation is accompanied by dynamic changes in histone acetylation of cytokine genes. *Nat Immunol* 3 (7), 643–51.

Bach, E. A., Szabo, S. J., Dighe, A. S., Ashkenazi, A., Aguet, M., Murphy, K. M., Schreiber, R. D. 1995. Ligand-induced autoregulation of IFN-gamma receptor beta chain expression in T helper cell subsets. *Science* 270 (5239), 1215–8.

Bacon, C. M., Petricoin, E. F., Ortaldo, J. R., Rees, R. C., Larner, A. C., Johnston, J. A., O'Shea, J. J. 1995. Interleukin 12 induces tyrosine phosphorylation and activation of STAT4 in human lymphocytes. *Proc Natl Acad Sci U S A* 92 (16), 7307–7311.

Baek, K.-H., Zaslavsky, A., Lynch, R. C., Britt, C., Okada, Y., Siarey, R. J., Lensch, M. W., Park, I.-H., Yoon, S. S., Minami, T., Korenberg, J. R., Folkman, J., Daley, G. Q., Aird, W. C., Galdzicki, Z., Ryeom, S. 2009. Down's syndrome suppression of tumour growth and the role of the calcineurin inhibitor DSCR1. *Nature* 459 (7250), 1126–1130.

Beals, C. R., Clipstone, N. A., Ho, S. N., Crabtree, G. R. 1997. Nuclear localization of NF-ATc by a calcineurin-dependent, cyclosporin-sensitive intramolecular interaction. *Genes Dev* 11 (7), 824–834.

Beals, C. R., Sheridan, C. M., Turck, C. W., Gardner, P., Crabtree, G. R. 1997. Nuclear export of NF-ATc enhanced by glycogen synthase kinase-3. *Science* 275 (5308), 1930–1934.

Beima, K. M., Miazgowicz, M. M., Lewis, M. D., Yan, P. S., Huang, T. H.-M., Weinmann, A. S. 2006. T-bet binding to newly identified target gene promoters is cell type-independent but results in variable context-dependent functional effects. *J Biol Chem* 281 (17), 11992–12000.

Bernabei, P., Allione, A., Rigamonti, L., Bosticardo, M., Losana, G., Borghi, I., Forni, G., Novelli, F. 2001. Regulation of interferon-gamma receptor chains: a peculiar way to rule the life and death of human lymphocytes. *Eur Cytokine Netw* 12 (1), 6–14.

Bird, J. J., Brown, D. R., Mullen, A. C., Moskowitz, N. H., Mahowald, M. A., Sider, J. R., Gajewski, T. F., Wang, C. R., Reiner, S. L. 1998. Helper T cell differentiation is controlled by the cell cycle. *Immunity* 9 (2), 229–37.

Busse, D., 2009. Dynamics of the IL-2 cytokine network and T-cell proliferation. Ph.D. thesis, Humboldt Universität.

Carpenter, A. E., Jones, T. R., Lamprecht, M. R., Clarke, C., Kang, I. H., Friman, O., Guertin, D. A., Chang, J. H., Lindquist, R. A., Moffat, J., Golland, P., Sabatini, D. M., 2006. CellProfiler: image analysis software for identifying and quantifying cell phenotypes. *Genome Biol* 7 (10), R100.

Cava, A. L. 2009. Lupus and T cells. *Lupus* 18 (3), 196–201.

Chang, H. D., Helbig, C., Tykocinski, L., Kreher, S., Koeck, J., Niesner, U., Radbruch, A. 2007. Expression of IL-10 in Th memory lymphocytes is conditional on IL-12 or IL-4, unless the IL-10 gene is imprinted by GATA-3. *Eur J Immunol* 37 (3), 807–17.

Chang, J. T., Shevach, E. M., Segal, B. M. 1999. Regulation of interleukin (IL)-12 receptor beta2 subunit expression by endogenous IL-12: a critical step in the differentiation of pathogenic autoreactive T cells. *J Exp Med* 189 (6), 969–78.

Chang, S., Aune, T. M. 2005. Histone hyperacetylated domains across the Ifng gene region in natural killer cells and T cells. *Proc Natl Acad Sci U S A* 102 (47), 17095–100.

Chang, S., Aune, T. M. 2007. Dynamic changes in histone-methylation 'marks' across the locus encoding interferon-gamma during the differentiation of T helper type 2 cells. *Nat Immunol* 8 (7), 723–31.

Corn, R. A., Hunter, C., Liou, H.-C., Siebenlist, U., Boothby, M. R. 2005. Opposing roles for RelB and Bcl-3 in regulation of T-box expressed in T cells, GATA-3, and Th effector differentiation. *J Immunol* 175 (4), 2102–2110.

Cross, D. A., Alessi, D. R., Cohen, P., Andjelkovich, M., Hemmings, B. A., 1995. Inhibition of glycogen synthase kinase-3 by insulin mediated by protein kinase B. *Nature* 378 (6559), 785–789.

Curotto de Lafaille, M. A., Lafaille, J. J. 2009. Natural and adaptive foxp3+ regulatory T cells: more of the same or a division of labor? *Immunity* 30 (5), 626–635.

Diehn, M., Alizadeh, A. A., Rando, O. J., Liu, C. L., Stankunas, K., Botstein, D., Crabtree, G. R., Brown, P. O. 2002. Genomic expression programs and the integration of the CD28 costimulatory signal in T cell activation. *Proc Natl Acad Sci U S A* 99 (18), 11796–11801.

Djuretic, I. M., Levanon, D., Negreanu, V., Groner, Y., Rao, A., Ansel, K. M. 2007. Transcription factors T-bet and Runx3 cooperate to activate Ifng and silence Il4 in T helper type 1 cells. *Nat Immunol* 8 (2), 145–53.

Efron, B., Tibshirani, R., 1993. *An Introduction to the bootstrap*. Chapman & Hall, New York.

Fields, P. E., Kim, S. T., Flavell, R. A. 2002. Cutting edge: changes in histone acetylation at the IL-4 and IFN-gamma loci accompany Th1/Th2 differentiation. *J Immunol* 169 (2), 647–50.

Glimcher, L. H., Murphy, K. M. 2000. Lineage commitment in the immune system: the T helper lymphocyte grows up. *Genes Dev* 14 (14), 1693–711.

Gómez del Arco, P., Martínez-Martínez, S., Maldonado, J. L., Ortega-Pérez, I., Redondo, J. M. 2000. A role for the p38 MAP kinase pathway in the nuclear shuttling of NFATp. *J Biol Chem* 275 (18), 13872–13878.

Gorelik, L., Constant, S., Flavell, R. A. 2002. Mechanism of transforming growth factor beta-induced inhibition of T helper type 1 differentiation. *J Exp Med* 195 (11), 1499–1505.

Grenningloh, R., Kang, B. Y., Ho, I. C. 2005. Ets-1, a functional cofactor of T-bet, is essential for Th1 inflammatory responses. *J Exp Med* 201 (4), 615–26.

Grogan, J. L., Mohrs, M., Harmon, B., Lacy, D. A., Sedat, J. W., Locksley, R. M. 2001. Early transcription and silencing of cytokine genes underlie polarization of T helper cell subsets. *Immunity* 14 (3), 205–15.

Gwack, Y., Sharma, S., Nardone, J., Tanasa, B., Iuga, A., Srikanth, S., Okamura, H., Bolton, D., Feske, S., Hogan, P. G., Rao, A. 2006. A genome-wide Drosophila RNAi screen identifies DYRK-family kinases as regulators of NFAT. *Nature* 441 (7093), 646–650.

Hartwell, L. H., Hopfield, J. J., Leibler, S., Murray, A. W. 1999. From molecular to modular cell biology. *Nature* 402 (6761 Suppl), C47–C52.

Hasty, J., McMillen, D., Isaacs, F., Collins, J. J. 2001. Computational studies of gene regulatory networks: in numero molecular biology. *Nat Rev Genet* 2 (4), 268–279.

Hatton, R. D., Harrington, L. E., Luther, R. J., Wakefield, T., Janowski, K. M., Oliver, J. R., Lallone, R. L., Murphy, K. M., Weaver, C. T. 2006. A distal conserved sequence element controls Ifng gene expression by T cells and NK cells. *Immunity* 25 (5), 717–29.

Heath, V. L., Showe, L., Crain, C., Barrat, F. J., Trinchieri, G., O'Garra, A. 2000. Cutting edge: ectopic expression of the IL-12 receptor-beta 2 in developing and committed Th2 cells does not affect the production of IL-4 or induce the production of IFN-gamma. *J Immunol* 164 (6), 2861–5.

Höfer, T., Nathansen, H., Löhning, M., Radbruch, A., Heinrich, R. 2002. GATA-3 transcriptional imprinting in Th2 lymphocytes: a mathematical model. *Proc Natl Acad Sci U S A* 99 (14), 9364–9368.

Himmelrich, H., Parra-Lopez, C., Tacchini-Cottier, F., Louis, J. A., Launois, P. 1998. The IL-4 rapidly produced in BALB/c mice after infection with Leishmania major down-regulates IL-12 receptor beta 2-chain expression on CD4+ T cells resulting in a state of unresponsiveness to IL-12. *J Immunol* 161 (11), 6156–6163.

Hogan, P. G., Chen, L., Nardone, J., Rao, A. 2003. Transcriptional regulation by calcium, calcineurin, and NFAT. *Genes Dev* 17 (18), 2205–2232.

Hsieh, C. S., Macatonia, S. E., Tripp, C. S., Wolf, S. F., O'Garra, A., Murphy, K. M. 1993. Development of TH1 CD4+ T cells through IL-12 produced by Listeria-induced macrophages. *Science* 260 (5107), 547–9.

Hu-Li, J., Huang, H., Ryan, J., Paul, W. E. 1997. In differentiated CD4+ T cells, interleukin 4 production is cytokine-autonomous, whereas interferon gamma production is cytokine-dependent. *Proc Natl Acad Sci U S A* 94 (7), 3189–94.

Hwang, E. S., Szabo, S. J., Schwartzberg, P. L., Glimcher, L. H. 2005. T helper cell fate specified by kinase-mediated interaction of T-bet with GATA-3. *Science* 307 (5708), 430–3.

Jacobson, N. G., Szabo, S. J., Weber-Nordt, R. M., Zhong, Z., Schreiber, R. D., Darnell, J. E., Murphy, K. M. 1995. Interleukin 12 signaling in T helper type 1 (Th1) cells involves tyrosine phosphorylation of signal transducer and activator of transcription (Stat)3 and Stat4. *J Exp Med* 181 (5), 1755–1762.

Jones, B., Chen, J. 2006. Inhibition of IFN-gamma transcription by site-specific methylation during T helper cell development. *Embo J* 25 (11), 2443–52.

Kano, S. I., Sato, K., Morishita, Y., Vollstedt, S., Kim, S., Bishop, K., Honda, K., Kubo, M., Taniguchi, T. 2008. The contribution of transcription factor IRF1 to the interferon-gamma-interleukin 12 signaling axis and T(H)1 versus T-H-17 differentiation of CD4(+) T cells. *Nature Immunology* 9 (1), 34–41.

Kaplan, M. H., Sun, Y. L., Hoey, T., Grusby, M. J. 1996. Impaired IL-12 responses and enhanced development of Th2 cells in Stat4-deficient mice. *Nature* 382 (6587), 174–177.

Kehlenbach, R. H., Dickmanns, A., Gerace, L. 1998. Nucleocytoplasmic shuttling factors including Ran and CRM1 mediate nuclear export of NFAT In vitro. *J Cell Biol* 141 (4), 863–874.

Kilka, S., Erdmann, F., Migdoll, A., Fischer, G., Weiwad, M. 2009. The proline-rich N-terminal sequence of calcineurin Abeta determines substrate binding. *Biochemistry* 48 (9), 1900–1910.

Kirkpatrick, S., Gelatt, C. D., J., Vecchi, M. P. 1983. Optimization by Simulated Annealing. *Science* 220 (4598), 671–680.

Koch, F., Stanzl, U., Jennewein, P., Janke, K., Heufler, C., Kämpgen, E., Romani, N., Schuler, G. 1996. High level IL-12 production by murine dendritic cells: upregulation via MHC class II and CD40 molecules and downregulation by IL-4 and IL-10. *J Exp Med* 184 (2), 741–746.

Lighvani, A. A., Frucht, D. M., Jankovic, D., Yamane, H., Aliberti, J., Hissong, B. D., Nguyen, B. V., Gadina, M., Sher, A., Paul, W. E., O'Shea, J. J. 2001. T-bet is rapidly induced by interferon-gamma in lymphoid and myeloid cells. *Proc Natl Acad Sci U S A* 98 (26), 15137–42.

Litman, G. W., Cannon, J. P., Dishaw, L. J. 2005. Reconstructing immune phylogeny: new perspectives. *Nat Rev Immunol* 5 (11), 866–879.

Liu, J., Farmer, J. D., Lane, W. S., Friedman, J., Weissman, I., Schreiber, S. L. 1991. Calcineurin is a common target of cyclophilin-cyclosporin A and FKBP-FK506 complexes. *Cell* 66 (4), 807–815.

Liu, J. O. 2009. Calmodulin-dependent phosphatase, kinases, and transcriptional corepressors involved in T-cell activation. *Immunol Rev* 228 (1), 184–198.

Liu, Y., Janeway, C. A. 1990. Interferon gamma plays a critical role in induced cell death of effector T cell: a possible third mechanism of self-tolerance. *J Exp Med* 172 (6), 1735–1739.

Loh, C., Carew, J. A., Kim, J., Hogan, P. G., Rao, A. 1996. T-cell receptor stimulation elicits an early phase of activation and a later phase of deactivation of the transcription factor NFAT1. *Mol Cell Biol* 16 (7), 3945–3954.

Macatonia, S. E., Hosken, N. A., Litton, M., Vieira, P., Hsieh, C. S., Culpepper, J. A., Wysocka, M., Trinchieri, G., Murphy, K. M., O'Garra, A. 1995. Dendritic cells produce IL-12 and direct the development of Th1 cells from naive CD4+ T cells. *J Immunol* 154 (10), 5071–9.

Macatonia, S. E., Hsieh, C. S., Murphy, K. M., O'Garra, A. 1993. Dendritic cells and macrophages are required for Th1 development of CD4+ T cells from alpha beta TCR transgenic mice: IL-12 substitution for macrophages to stimulate IFN-gamma production is IFN-gamma-dependent. *Int Immunol* 5 (9), 1119–28.

Magram, J., Connaughton, S. E., Warrier, R. R., Carvajal, D. M., Wu, C. Y., Ferrante, J., Stewart, C., Sarmiento, U., Faherty, D. A., Gately, M. K. 1996. IL-12-deficient mice are defective in IFN gamma production and type 1 cytokine responses. *Immunity* 4 (5), 471–481.

Manke, T., Roider, H. G., Vingron, M. 2008. Statistical modeling of transcription factor binding affinities predicts regulatory interactions. *PLoS Comput Biol* 4 (3), e1000039.

Mariani, L., Löhning, M., Radbruch, A., Höfer, T. 2004. Transcriptional control networks of cell differentiation: insights from helper T lymphocytes. *Prog Biophys Mol Biol* 86 (1), 45–76.

Martins, G. A., Hutchins, A. S., Reiner, S. L. 2005. Transcriptional activators of helper T cell fate are required for establishment but not maintenance of signature cytokine expression. *J Immunol* 175 (9), 5981–5.

McGeachy, M. J., Cua, D. J. 2008. Th17 cell differentiation: the long and winding road. *Immunity* 28 (4), 445–453.

Mendoza, L. 2006. A network model for the control of the differentiation process in Th cells. *Biosystems* 84 (2), 101–14.

Mikhalkevich, N., Becknell, B., Caligiuri, M. A., Bates, M. D., Harvey, R., Zheng, W. P. 2006. Responsiveness of naive CD4 T cells to polarizing cytokine determines the ratio of Th1 and Th2 cell differentiation. *J Immunol* 176 (3), 1553–60.

Miller, S. A., Huang, A. C., Miazgowicz, M. M., Brassil, M. M., Weinmann, A. S. 2008. Coordinated but physically separable interaction with H3K27-demethylase and H3K4-methyltransferase activities are required for T-box protein-mediated activation of developmental gene expression. *Genes Dev* 22 (21), 2980–2993.

Minter, L. M., Turley, D. M., Das, P., Shin, H. M., Joshi, I., Lawlor, R. G., Cho, O. H., Palaga, T., Gottipati, S., Telfer, J. C., Kostura, L., Fauq, A. H., Simpson, K., Such, K. A., Miele, L., Golde, T. E., Miller, S. D., Osborne, B. A. 2005. Inhibitors of gamma-secretase block in vivo and in vitro T helper type 1 polarization by preventing Notch upregulation of Tbx21. *Nat Immunol* 6 (7), 680–8.

Mosser, D. M., Edwards, J. P. 2008. Exploring the full spectrum of macrophage activation. *Nat Rev Immunol* 8 (12), 958–969.

Mullen, A. C., High, F. A., Hutchins, A. S., Lee, H. W., Villarino, A. V., Livingston, D. M., Kung, A. L., Cereb, N., Yao, T. P., Yang, S. Y., Reiner, S. L. 2001. Role of T-bet in commitment of TH1 cells before IL-12-dependent selection. *Science* 292 (5523), 1907–10.

Mullen, A. C., Hutchins, A. S., High, F. A., Lee, H. W., Sykes, K. J., Chodosh, L. A., Reiner, S. L. 2002. Hlx is induced by and genetically interacts with T-bet to promote heritable T(H)1 gene induction. *Nat Immunol* 3 (7), 652–8.

Murphy, K. M. 2006. Stress management for T helper differentiation. *Nat Immunol* 7 (6), 553–5.

Murphy, K. M., Reiner, S. L. 2002. The lineage decisions of helper T cells. *Nat Rev Immunol* 2 (12), 933–44.

Murphy, K. M., Tranvers, P., Walport, M., 2008. *Janeway's Immunobiology*. Taylor & Francis.

Naoe, Y., Setoguchi, R., Akiyama, K., Muroi, S., Kuroda, M., Hatam, F., Littman, D. R., Taniuchi, I. 2007. Repression of interleukin-4 in T helper type 1 cells by Runx/Cbf beta binding to the Il4 silencer. *Journal of Experimental Medicine* 204 (8), 1749–1755.

Neal, J. W., Clipstone, N. A. 2001. Glycogen synthase kinase-3 inhibits the DNA binding activity of NFATc. *J Biol Chem* 276 (5), 3666–3673.

Nishikomori, R., Ehrhardt, R. O., Strober, W. 2000. T helper type 2 cell differentiation occurs in the presence of interleukin 12 receptor beta2 chain expression and signaling. *J Exp Med* 191 (5), 847–58.

O'Garra, A., Vieira, P. 2007. T(H)1 cells control themselves by producing interleukin-10. *Nat Rev Immunol* 7 (6), 425–8.

Okamura, H., Aramburu, J., García-Rodríguez, C., Viola, J. P., Raghavan, A., Tahiliani, M., Zhang, X., Qin, J., Hogan, P. G., Rao, A. 2000. Concerted dephosphorylation of the transcription factor NFAT1 induces a conformational switch that regulates transcriptional activity. *Mol Cell* 6 (3), 539–550.

Okamura, H., Garcia-Rodriguez, C., Martinson, H., Qin, J., Virshup, D. M., Rao, A. 2004. A conserved docking motif for CK1 binding controls the nuclear localization of NFAT1. *Mol Cell Biol* 24 (10), 4184–4195.

Orange, J. S., Wang, B., Terhorst, C., Biron, C. A. 1995. Requirement for natural killer cell-produced interferon gamma in defense against murine cytomegalovirus infection and enhancement of this defense pathway by interleukin 12 administration. *J Exp Med* 182 (4), 1045–56.

Ouyang, W., Jacobson, N. G., Bhattacharya, D., Gorham, J. D., Fenoglio, D., Sha, W. C., Murphy, T. L., Murphy, K. M. 1999. The Ets transcription factor ERM is Th1-specific and induced by IL-12 through a Stat4-dependent pathway. *Proc Natl Acad Sci U S A* 96 (7), 3888–3893.

Ouyang, W., Ranganath, S. H., Weindel, K., Bhattacharya, D., Murphy, T. L., Sha, W. C., Murphy, K. M. 1998. Inhibition of Th1 development mediated by GATA-3 through an IL-4-independent mechanism. *Immunity* 9 (5), 745–55.

Pernis, A., Gupta, S., Gollob, K. J., Garfein, E., Coffman, R. L., Schindler, C., Rothman, P. 1995. Lack of interferon gamma receptor beta chain and the prevention of interferon gamma signaling in TH1 cells. *Science* 269 (5221), 245–247.

Perussia, B., Chan, S. H., D'Andrea, A., Tsuji, K., Santoli, D., Pospisil, M., Young, D., Wolf, S. F., Trinchieri, G. 1992. Natural killer (NK) cell stimulatory factor or IL-12 has differential effects on the proliferation of TCR-alpha beta+, TCR-gamma delta+ T lymphocytes, and NK cells. *J Immunol* 149 (11), 3495–3502.

Raue, A., Kreutz, C., Maiwald, T., Bachmann, J., Schilling, M., Klingmüller, U., Timmer, J. 2009. Structural and practical identifiability analysis of partially observed dynamical models by exploiting the profile likelihood. *Bioinformatics* 25 (15), 1923–1929.

Richter, A., Lohning, M., Radbruch, A. 1999. Instruction for cytokine expression in T helper lymphocytes in relation to proliferation and cell cycle progression. *J Exp Med* 190 (10), 1439–50.

Rogge, L., Barberis-Maino, L., Biffi, M., Passini, N., Presky, D. H., Gubler, U., Sinigaglia, F. 1997. Selective expression of an interleukin-12 receptor component by human T helper 1 cells. *J Exp Med* 185 (5), 825–31.

Romagnani, S. 1996. Th1 and Th2 in human diseases. *Clin Immunol Immunopathol* 80 (3 Pt 1), 225–35.

Rutz, S., Janke, M., Kassner, N., Hohnstein, T., Krueger, M., Scheffold, A. 2008. Notch regulates IL-10 production by T helper 1 cells. *Proc Natl Acad Sci U S A* 105 (9), 3497–502.

Salazar, C., Höfer, T. 2003. Allosteric regulation of the transcription factor NFAT1 by multiple phosphorylation sites: a mathematical analysis. *J Mol Biol* 327 (1), 31–45.

Scharton, T. M., Scott, P. 1993. Natural killer cells are a source of interferon gamma that drives differentiation of CD4+ T cell subsets and induces early resistance to Leishmania major in mice. *J Exp Med* 178 (2), 567–577.

Schmitt, E., Hoehn, P., Huels, C., Goedert, S., Palm, N., Rude, E., Germann, T. 1994. T helper type 1 development of naive CD4+ T cells requires the coordinate action of interleukin-12 and interferon-gamma and is inhibited by transforming growth factor-beta. *Eur J Immunol* 24 (4), 793–8.

Schroder, K., Hertzog, P. J., Ravasi, T., Hume, D. A. 2004. Interferon-gamma: an overview of signals, mechanisms and functions. *J Leukoc Biol* 75 (2), 163–189.

Scott, P. 1991. IFN-gamma modulates the early development of Th1 and Th2 responses in a murine model of cutaneous leishmaniasis. *J Immunol* 147 (9), 3149–55.

Seder, R. A., Gazzinelli, R., Sher, A., Paul, W. E. 1993. Interleukin 12 acts directly on CD4+ T cells to enhance priming for interferon gamma production and diminishes interleukin 4 inhibition of such priming. *Proc Natl Acad Sci U S A* 90 (21), 10188–92.

Shaw, K. T., Ho, A. M., Raghavan, A., Kim, J., Jain, J., Park, J., Sharma, S., Rao, A., Hogan, P. G. 1995. Immunosuppressive drugs prevent a rapid dephosphorylation of transcription factor NFAT1 in stimulated immune cells. *Proc Natl Acad Sci U S A* 92 (24), 11205–11209.

Sheridan, C. M., Heist, E. K., Beals, C. R., Crabtree, G. R., Gardner, P. 2002. Protein kinase A negatively modulates the nuclear accumulation of NF-ATc1 by priming for subsequent phosphorylation by glycogen synthase kinase-3. *J Biol Chem* 277 (50), 48664–48676.

Shi, M., Lin, T. H., Appell, K. C., Berg, L. J. 2008. Janus-kinase-3-dependent signals induce chromatin remodeling at the Ifng locus during T helper 1 cell differentiation. *Immunity* 28 (6), 763–73.

Shnyreva, M., Weaver, W. M., Blanchette, M., Taylor, S. L., Tompa, M., Fitzpatrick, D. R., Wilson, C. B. 2004. Evolutionarily conserved sequence elements that positively regulate IFN-gamma expression in T cells. *Proc Natl Acad Sci U S A* 101 (34), 12622–7.

Shuai, K., Liu, B. 2003. Regulation of JAK-STAT signalling in the immune system. *Nat Rev Immunol* 3 (11), 900–911.

Sieber, M., Karanik, M., Brandt, C., Blex, C., Podtschaske, M., Erdmann, F., Rost, R., Serfling, E., Liebscher, J., Patzel, M., Radbruch, A., Fischer, G., Baumgrass, R. 2007. Inhibition of calcineurin-NFAT signaling by the pyrazolopyrimidine compound NCI3. *Eur J Immunol* 37 (9), 2617–26.

Smeltz, R. B., Chen, J., Ehrhardt, R., Shevach, E. M. 2002. Role of IFN-gamma in Th1 differentiation: IFN-gamma regulates IL-18R alpha expression by preventing the negative effects of IL-4 and by inducing/maintaining IL-12 receptor beta 2 expression. *J Immunol* 168 (12), 6165–72.

Stemmer, P. M., Klee, C. B. 1994. Dual calcium ion regulation of calcineurin by calmodulin and calcineurin B. *Biochemistry* 33 (22), 6859–6866.

Stobie, L., Gurunathan, S., Prussin, C., Sacks, D. L., Glaichenhaus, N., Wu, C. Y., Seder, R. A. 2000. The role of antigen and IL-12 in sustaining Th1 memory cells in vivo: IL-12 is required to maintain memory/effector Th1 cells sufficient to mediate protection to an infectious parasite challenge. *Proc Natl Acad Sci U S A* 97 (15), 8427–8432.

Szabo, S. J., Dighe, A. S., Gubler, U., Murphy, K. M. 1997. Regulation of the interleukin (IL)-12R beta 2 subunit expression in developing T helper 1 (Th1) and Th2 cells. *J Exp Med* 185 (5), 817–24.

Szabo, S. J., Kim, S. T., Costa, G. L., Zhang, X., Fathman, C. G., Glimcher, L. H. 2000. A novel transcription factor, T-bet, directs Th1 lineage commitment. *Cell* 100 (6), 655–69.

Szabo, S. J., Sullivan, B. M., Stemmann, C., Satoskar, A. R., Sleckman, B. P., Glimcher, L. H. 2002. Distinct effects of T-bet in TH1 lineage commitment and IFN-gamma production in CD4 and CD8 T cells. *Science* 295 (5553), 338–42.

Thierfelder, W. E., van Deursen, J. M., Yamamoto, K., Tripp, R. A., Sarawar, S. R., Carson, R. T., Sangster, M. Y., Vignali, D. A., Doherty, P. C., Grosveld, G. C., Ihle, J. N. 1996. Requirement for Stat4 in interleukin-12-mediated responses of natural killer and T cells. *Nature* 382 (6587), 171–174.

Thieu, V. T., Yu, Q., Chang, H. C., Yeh, N., Nguyen, E. T., Sehra, S., Kaplan, M. H. 2008. Signal transducer and activator of transcription 4 is required for the transcription factor T-bet to promote T helper 1 cell-fate determination. *Immunity* 29 (5), 679–90.

Timmer, J., Müller, T., Swameye, I., Sandra, O., Klingmüller, U., 2004. Modelling the nonlinear dynamics of cellular signal transduction. *Int. J. Bif. Chaos* 14, 2069–2079.

Timmerman, L. A., Clipstone, N. A., Ho, S. N., Northrop, J. P., Crabtree, G. R. 1996. Rapid shuttling of NF-AT in discrimination of Ca2+ signals and immunosuppression. *Nature* 383 (6603), 837–840.

Tomlin, C. J., Axelrod, J. D. 2007. Biology by numbers: mathematical modelling in developmental biology. *Nat Rev Genet* 8 (5), 331–340.

Tong, Y., Aune, T., Boothby, M. 2005. T-bet antagonizes mSin3a recruitment and transactivates a fully methylated IFN-gamma promoter via a conserved T-box half-site. *Proc Natl Acad Sci U S A* 102 (6), 2034–9.

Trinchieri, G. 2003. Interleukin-12 and the regulation of innate resistance and adaptive immunity. *Nat Rev Immunol* 3 (2), 133–46.

Usui, T., Nishikomori, R., Kitani, A., Strober, W. 2003. GATA-3 suppresses Th1 development by downregulation of Stat4 and not through effects on IL-12Rbeta2 chain or T-bet. *Immunity* 18 (3), 415–28.

Usui, T., Preiss, J. C., Kanno, Y., Yao, Z. J., Bream, J. H., O'Shea J, J., Strober, W. 2006. T-bet regulates Th1 responses through essential effects on GATA-3 function rather than on IFNG gene acetylation and transcription. *J Exp Med.*

van Rietschoten, J. G., Smits, H. H., van de Wetering, D., Westland, R., Verweij, C. L., den Hartog, M. T., Wierenga, E. A. 2001. Silencer activity of NFATc2 in the interleukin-12 receptor beta 2 proximal promoter in human T helper cells. *J Biol Chem* 276 (37), 34509–16.

Wakil, A. E., Wang, Z. E., Ryan, J. C., Fowell, D. J., Locksley, R. M. 1998. Interferon gamma derived from CD4(+) T cells is sufficient to mediate T helper cell type 1 development. *J Exp Med* 188 (9), 1651–6.

Watford, W. T., Hissong, B. D., Bream, J. H., Kanno, Y., Muul, L., O'Shea, J. J. 2004. Signaling by IL-12 and IL-23 and the immunoregulatory roles of STAT4. *Immunol Rev* 202, 139–56.

Wei, G., Wei, L., Zhu, J., Zang, C., Hu-Li, J., Yao, Z., Cui, K., Kanno, Y., Roh, T.-Y., Watford, W. T., Schones, D. E., Peng, W., Sun, H.-W., Paul, W. E., O'Shea, J. J., Zhao, K. 2009. Global mapping of H3K4me3 and H3K27me3 reveals specificity and plasticity in lineage fate determination of differentiating CD4+ T cells. *Immunity* 30 (1), 155–167.

Wenner, C. A., Guler, M. L., Macatonia, S. E., O'Garra, A., Murphy, K. M. 1996. Roles of IFN-gamma and IFN-alpha in IL-12-induced T helper cell-1 development. *J Immunol* 156 (4), 1442–7.

Wilson, C. B., Rowell, E., Sekimata, M. 2009. Epigenetic control of T-helper-cell differentiation. *Nat Rev Immunol* 9 (2), 91–105.

Yamane, H., Igarashi, O., Kato, T., Nariuchi, H. 2000. Positive and negative regulation of IL-12 receptor expression of naive CD4(+) T cells by CD28/CD152 co-stimulation. *Eur J Immunol* 30 (11), 3171–3180.

Yang, Y., Ochando, J. C., Bromberg, J. S., Ding, Y. Z. 2007. Of a distant T-bet enhancer responsive to IL-12/Stat4 and IFN gamma/Stat1 signals. *Blood* 110 (7), 2494–2500.

Yap, G., Pesin, M., Sher, A. 2000. Cutting edge: IL-12 is required for the maintenance of IFN-gamma production in T cells mediating chronic resistance to the intracellular pathogen, Toxoplasma gondii. *J Immunol* 165 (2), 628–31.

Yates, A., Callard, R., Stark, J. 2004. Combining cytokine signalling with T-bet and GATA-3 regulation in Th1 and Th2 differentiation: a model for cellular decision-making. *J Theor Biol* 231 (2), 181–96.

Yoshikai, Y., 2006. Immunological protection against mycobacterium tuberculosis infection. *Crit Rev Immunol* 26 (6), 515–526.

Zhang, F., Boothby, M. 2006. T helper type 1-specific Brg1 recruitment and remodeling of nucleosomes positioned at the IFN-gamma promoter are Stat4 dependent. *J Exp Med.*

Zhang, S., Zhang, H., Zhao, J. 2009. The role of CD4 T cell help for CD8 CTL activation. *Biochem Biophys Res Commun* 384 (4), 405–408.

Zhou, L., Chong, M. M. W., Littman, D. R. 2009. Plasticity of CD4+ T cell lineage differentiation. *Immunity* 30 (5), 646–655.

Zhu, J., McKeon, F. 1999. NF-AT activation requires suppression of Crm1-dependent export by calcineurin. *Nature* 398 (6724), 256–260.

Zhu, J., Paul, W. E. 2008. CD4 T cells: fates, functions, and faults. *Blood* 112 (5), 1557–69.

Zhu, J., Shibasaki, F., Price, R., Guillemot, J. C., Yano, T., Dötsch, V., Wagner, G., Ferrara, P., McKeon, F. 1998. Intramolecular masking of nuclear import signal on NF-AT4 by casein kinase I and MEKK1. *Cell* 93 (5), 851–861.

Danksagung

Zuallererst möchte ich meinen Betreuern Thomas Höfer und Andreas Radbruch dafür danken, dass sie mir die Möglichkeit eröffnet haben, dieses interdisziplinäre Projekt zu bearbeiten. Bei Thomas möchte ich mich insbesondere dafür bedanken, dass er mich in die Kunst der mathematischen Modellierung eingewiesen hat. Andreas danke ich dafür, dass er mir beigebracht hat, wie man ein gutes Paper schreibt und dann auch noch Editorin und Reviewer davon überzeugt.

Dafür, dass sie ihr Wissen und ihre Erfahrung mit mir geteilt haben, möchte ich mich bei den Mitgliedern der AG Höfer und der AG Radbruch bedanken. Mein größter Dank gilt natürlich Luca, der mir alles beigebracht hat, was ich jetzt über Modellierung weiß. Außerdem war es immer wieder toll mit ihm zusammenzuarbeiten, wegen seiner Kreativität, seinem Abstraktionsvermögen und natürlich seinem Enthusiasmus. Immerhin hat Luca zuerst die zwei Loops entdeckt! Als die AG Höfer noch in Berlin beheimatet war, habe ich sehr von der Hilfsbereitschaft aller profitiert. In diesem Zusammenhang, möchte ich mich insbesondere bei Stephan und Toni bedanken. Später in unserer Exilgruppenzeit, fand ich es sehr schön, alle Aufs und Abs mit Dorothea teilen zu können. Natürlich möchte ich mich beim ITB bedanken, dass wir Heimatlose so freundlich aufgenommen wurden.

Auch am DRFZ wusste ich die hilfsbereite Atmosphäre immer sehr zu schätzen. Insbesondere haben Uwe und Inka mich mit unseren T-Zellen vertraut gemacht und Maria hat ihre Erfahrungen mit Transkriptionsfaktor-Färbung und Secretion-Assay mit mir geteilt. Und natürlich möchte ich mich für die lustigen Abende auf der Dachterasse und in der Raucherecke bedanken. Auch Inga, Sandra und Guido möchte ich für ihre Hilfsbereitschaft im Labor danken. Koji, Hyun-Dong und Sascha möchte ich für ihre kritischen Kommentare während der Entstehung des Immunity-Paper danken. In diesem Zusammenhang auch ein Danke an Mark Ansel und Zoya.

For the really interesting experience in Boston, I want to thank Anjana Rao. It was a great opportunity to get to know such a place from inside. Also to see that some of the big shots are still mostly interested in the science was very motivating.

Zuletzt möchte ich noch ganz besonder Nils danken, für die tausend Dinge, die er auch fachlich zu meiner Arbeit beigetragen hat. Um nur Einiges zu nennen, die ganzen statistischen Methoden, die er mir beigebracht hat, die vielen Prozessoren seiner Supercomputer, ohne die ich die Paper-Revision nie rechtzeitig fertig gekriegt hätte. Und natürlich für die mentale Unterstützung, für sein Vertrauen, dass ich alles schaffen kann.

Als allerletztes kommen jetzt noch meine Eltern an die Reihe. Wenn sie mir nicht beigebracht hätten, dass man immer seine Meinung vertreten soll und v. a. seiner eigenen Meinung vertrauen soll, wäre diese Projekt wahrscheinlich nicht so gut gelaufen. Danke.

Zusammenfassung

Das Immunsystem der Säugetiere ist ein komplexes Netzwerk, in dem unterschiedliche Zelltypen zusammenarbeiten, um den Organismus vor fremden Eindringlingen zu schützen. T-Helfer-Lymphozyten spielen dabei eine zentrale Rolle, indem sie Signale erkennen, die Informationen über die Art des Eindringlings enthalten. In Abhängigkeit davon, in welcher Kombination diese Signale auftreten, wird ein Differenzierungsprozess ausgelöst, bei dem ein bestimmter Effektorzelltyp entsteht. Jeder Effektorzelltyp kann bestimmte Teile des Immunsystems aktivieren, die den Eindringling dann bekämpfen. In der vorliegenden Arbeit, wurde der Differenzierungprozess von naiven T-Helfer-Zellen zu Typ I (Th1) Zellen untersucht. Th1-Zellen lösen eine zellvermittelte Immunantwort aus durch Ausschüttung ihres Effektorzytokins Interferon-γ (IFN-γ). Th1-Zellen spielen eine wichtige Rolle bei der Immunabwehr intrazellulärer Erreger, ihre fehlgeleitete Aktivierung trägt aber auch maßgeblich zu vielen Autoimmunerkrankungen bei. Naive Th-Zellen differenzieren zu Th1-Zellen, wenn sie gleichzeitig mit der Stimulation ihres T-Zell-Rezeptors (TCR) das Zytokinsignal Interleukin-12 (IL-12) wahrnehmen. Welche Mechanismen der Wirkung von IL-12 zugrunde liegen, war bisher nur unvollständig verstanden. Es war bekannt, dass IL-12-Stimulation die Produktion eines weiteren Zytokins, IFN-γ, in den T-Zellen auslöst. Dieses induziert Expression des Transkriptionsfaktors T-bet, der den Differenzierungsprozess kontrolliert. Paradoxerweise genügt Stimulation durch IFN-γ ohne IL-12 jedoch nicht, um Th1-Differenzierung auszulösen, obwohl T-bet hochreguliert wird. Dieses seit langem ungeklärte Phänomen wurde in der vorliegenden Arbeit aufgeklärt.

Im ersten Teil der Arbeit wurde untersucht, welche Rolle IL-12 und IFN-γ in der Th1-Differenzierung spielen. Mathematische Modelle in Verbindung mit quantitativen Messungen wurden eingesetzt, um das gen-regulatorische Netzwerk zu untersuchen, das die Wirkung dieser Signale in T-Zellen vermittelt. Ausgehend von publizierten Erkenntnissen wurde ein mathematisches Modell des transkriptionellen Netzwerks entwickelt, das die Expression von drei wichtigen Th1-spezifischen Genen kontrolliert, nämlich IFN-γ, T-bet und die IL-12-Rezeptor β2-Kette (IL-12Rβ2), die die Empfindlichkeit von T-Zellen für IL-12 reguliert. Zur Überprüfung des Modells wurde die Expressionskinetik von T-bet, IFN-γ und IL-12Rβ2 während *in vitro* Th1-

Differenzierung von naiven primären Th-Zellen aus Mäusen gemessen. Da das literaturbasierte Modell die Daten nicht erklären konnte, wurde gezielt nach bisher unbekannten Regulationsmechanismen gesucht. Dabei wurden zwei bisher unbekannte regulatorische Wechselwirkungen entdeckt: IL-12, vermittelt durch STAT4, induziert T-bet-Expression direkt, allerding nur in der späten Phase der Differenzierung (>72 Std.). Dies konnte darauf zurückgeführt werden, dass die IL-12-Signalvermittlung in der frühen Phase blockiert ist, weil TCR-Signale die Expression des IL-12Rβ2 reprimieren. Durch Einbeziehung der entdeckten Regulationsmechanismen in das mathematische Modell konnten alle wesentlichen Merkmale der Expressionskinetiken reproduziert werden.

Die Analyse des vervollständigten Modells zeigte, dass T-bet in der frühen Phase (<48 Std.) durch eine bereits bekannte autokrine positive Rückkopplungsschleife mit IFN-γ reguliert wird. In der späten Phase induziert dann IL-12 T-bet-Expression. Dadurch wird eine weitere Rückkopplungsschleife aktiviert, in der T-bet die Expression von IL-12 induziert, was wiederum die IL-12-abhängige T-bet-Expression verstärkt. Im letzen Schritt wurde untersucht, inwieweit das T-bet-Expressionsniveau mit der Differenzierung korreliert. Dabei konnte gezeigt werden, dass T-bet nur in der späten Phase Th1-Differenzierung auslösen kann. Da T-bet-Expression in der späten Phase IL-12-abhängig ist, erklären die Ergebnisse, warum IL-12, aber nicht IFN-γ den Th1-Differenzierungprozess einleiten kann.

Im zweiten Teil der Arbeit wurde ein Netzwerk genauer untersucht, das den Transkriptionsfaktor NFAT reguliert und an der intrazellulären Signalverarbeitung nach Stimulation des TCR beteiligt ist. Die Aktivität von NFAT wird durch Dephosphorylierung mehrerer Serinreste reguliert. Die Phosphatase Calcineurin katalysiert diese Dephosphorylierung und löst dadurch Import des NFAT-Proteins in den Zellkern aus. Mehrere Kinasen, u. a. CK1 und GSK-3, sind an der Rephosphorylierung und der damit einhergehenden Deaktivierung beteiligt. Um die Rolle der Kinasen zu untersuchen, wurde ein mathematisches Modell des zugrunde liegenden Netzwerks entwickelt. Der Phosphorylierungszustand von NFAT und seine subzelluläre Verteilung wurden während der Aktivierungs- und Deaktivierungsphasen gemessen und einzelne Kinasen pharmakologisch gehemmt. Es wurde ein mathematisches Modell entwickelt, das Phosphorylierungs- und Transportprozesse beschreibt, und dessen kinetische Parameter aus den Daten abgeschätzt wurden. Die Modellanalyse ergab, dass die Phosphorylierung von NFAT stark kooperativ verlief. Außerdem konnte gezeigt werden, dass die Kinase GSK-3 eine wichtigere Rolle als CK1 bei der Regulation von NFAT spielt.

In der vorliegenden Arbeit wurden zwei Fragen der molekularen Immunologie bearbeitet. Dabei trug eine enge Vernetzung von Experiment und Theorie maßgeblich zum Erfolg der Arbeit bei.